Hamlyn all-colour paperbacks

KU-406-182

Paul Lewis BSc MB MRCP &
David Rubenstein MB MRCP

The Human Body

illustrated by John Bavosi

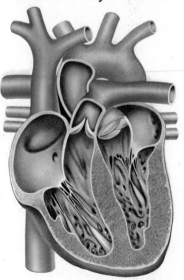

Hamlyn
London · New York · Sydney · Toronto

FOREWORD

The internal workings of the human body are something of a mystery to many people. Yet surely, for us, this must be the most fascinating organism of all.

This introduction to the body, its structure and functions, has been written and illustrated for those who wish to understand the basic systems of the body, how they work and how they are inter-related. Explanations are as detailed as space permits, without employing any more medical terminology than necessary. In a few sections numerical measurements of such things as blood pressure and partial pressures in respiration have been included for the more technically minded – but these can be skipped by the less technically minded as the basic principles are always clearly described.

It is hoped that this book will serve as a concise account of the subject both for those who are learning for their own interest and for those who will go on to study the subject in greater detail as part of a course.

Published by The Hamlyn Publishing Group Limited
London · New York · Sydney · Toronto
Astronaut House, Feltham, Middlesex, England

Copyright © The Hamlyn Publishing Group Limited 1970
Reprinted 1972, 1974, 1975
ISBN 0 600 00095 8

Phototypeset by Filmtype Services Limited, Scarborough, England
Colour separations by Schwitter Limited, Zurich
Printed in Spain by Mateu Cromo, Madrid

CONTENTS

CELLS AND TISSUES

Cells are the units from which all but the simplest living organisms are built. Although they range enormously in size and shape, all cells are essentially the same, and many cells in the human body closely resemble those found in primitive animals. Nevertheless most cells are specialized; grouped together with cells of similar structure and function they form tissues.

Nearly all human cells are invisible without high magnification. Their average diameter is between 1/50 and 1/100 mm, but they may be as small as 1/200 mm (certain brain cells) or as large as 1/4 mm (ova). When they are separate and surrounded by fluid, they are often spherical – as seen in the blood, and in cells cultured artificially in laboratories – but in solid tissues the pressures of adjoining cells modify this shape. Another reason for variation in size and shape is growth. Some cells grow as elongated cylinders or produce long shoots or branches.

The contents of a cell are the nucleus and the cytoplasm, which are enclosed within the cell wall, or membrane. This is extremely thin, being merely a double layer of lipid (fatty) molecules sandwiched between protein layers, and it restricts

Internal structure of a cell   (Opposite) Different types of cells

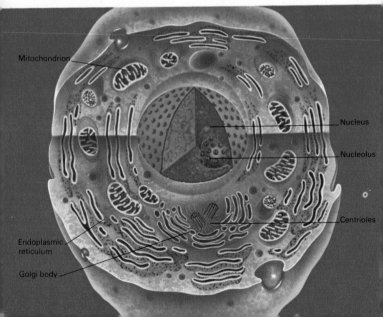

Mitochondrion

Nucleus

Nucleolus

Centrioles

Endoplasmic reticulum

Golgi body

the passage of materials in and out of the cell. The *nucleus* is rounded and contains the nucleolus and particles of chromatin, which are composed of DNA (deoxyribonucleic acid). The *cytoplasm* is a jelly-like solution of salts, proteins, lipids and carbohydrates. Some of its constituents are in minute granules and droplets; in addition it contains a number of more complex structures, the *organelles*. *Mitochondria* are spheres or tubules containing enzymes, the protein catalysts of biochemical reactions. These enzymes take part in respiration and energy release. The *endoplasmic reticulum* is a network of RNA (ribonucleic acid), and is the part of the cell which manufactures protein. Different RNA molecules have structures which are used like paper patterns to make different proteins over and over again. The *Golgi body* is prominent in cells that produce secretions and is a collection of fine channels close to the nucleus. Another organelle, the *centrosome* – also situated near the nucleus – comprises two centrioles surrounded by radiating filaments of cytoplasm. The centrosome is involved in cell division.

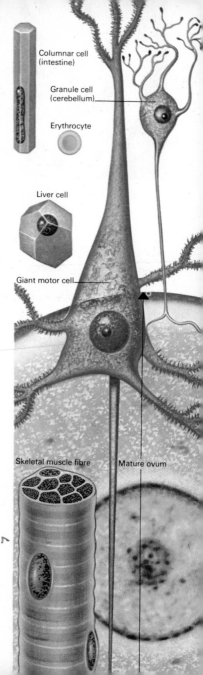

Columnar cell (intestine)

Granule cell (cerebellum)

Erythrocyte

Liver cell

Giant motor cell

Skeletal muscle fibre

Mature ovum

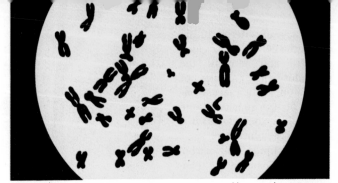
Human chromosomes

How cells divide

Nuclei are necessary for the maintenance and continuation of life. In growing tissues new cells are formed, and in almost every tissue, cells wear out and are replaced. New cell formation occurs through cell division. This is controlled by nuclei, which contain the messengers of heredity, the *genes*; genes are DNA molecules, and their duplication is the essential event when cells divide.

Genes transmit information about protein synthesis and enzyme manufacture, and thus they govern all vital processes. The sub-units of the DNA molecule are arranged in different ways in genes, enabling a vast quantity of information to be carried by a minute amount of material.

Genes are strung into long thin chains (chromatids or 'chromosomes') which are invisible in the non-dividing nucleus. There are forty-six chains (twenty-three pairs) in human cells. *Mitosis* is the usual type of division and is illustrated on the opposite page.

The formation of new DNA takes about eight hours, and mitosis lasts less than two hours. Even the fastest reproducing human cells divide no more often than daily, while some take weeks or months. Usually, therefore, division occurs as brief bursts of activity occupying only a small part of a cell's life-span and producing little disturbance of normal function.

Meiosis is a special type of division occurring in the ovaries and testes. It is sometimes known as reduction division, for the number of chromosomes in the cell progeny is reduced to half (twenty-three).

1 Early interphase. Chromatids are single. 2 Late interphase. Chromatids double to form identical chains attached to the parent chromatids only at their central points.

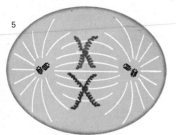
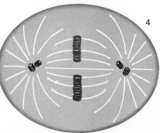

3 Prophase. Chains coil up. Centrosome splits into two parts which go to opposite ends of nucleus, remaining linked by spindle of rays of cytoplasm. Nucleus boundary disappears. 4 Early metaphase. Chromosomes are lined up and attached to spindle at central points.

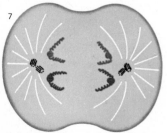
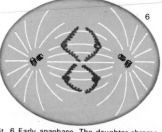

5 Late metaphase. Double chains begin to split. 6 Early anaphase. The daughter chromosomes part, as if drawn by the spindle.

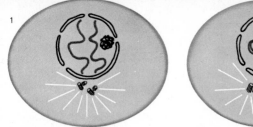
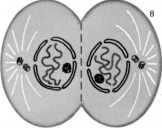

7 Late anaphase. 8 Telophase. Chromosomes elongate. Nucleus boundaries appear. There are now two cells, each with 46 chains of genes like its parent.

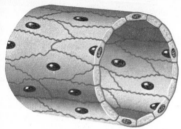

Simple epithelium lining arteries

Columnar epithelium as found in alimentary tract

Ciliated columnar epithelium as found in nostrils

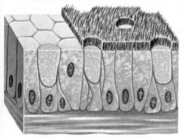

Transitional epithelium as found in bladder

Epithelial tissues

Epithelia are the sheets of tissue which cover and protect all external and internal surfaces of the body. They consist essentially of layers of cells packed together and lying on top of connective tissue. There are many different varieties of epithelium, from thick tough skin to the delicate lining of the lung alveoli, and with this diversity of structure goes a wide range of functions.

The simplest epithelia are formed of a single layer of flattened cells resembling a tiled pavement. Such cells are found in many places, including the tubules of the kidney, the inside of the eardrum, and the lungs, while similar cells line blood vessels and the pleural and peritoneal cavities. They are thin enough to allow the transfer of materials through them, while in

Stratified squamous epithelium as found in skin

the pleura and peritoneum they provide frictionless surfaces. The lining of the alimentary tract – from lower oesophagus to rectum – is much thicker, being composed of a layer of taller cells which under the microscope look like a row of books from the side and a honeycomb from on top. These cells secrete enzymes and mucus. A similar layer is found in the breathing passages, but here the free surface is covered with many fine hairs or *cilia*. Columnar epithelium of these sorts lines the ducts of some glands; other ducts have flatter cubical cells resembling the secretory cells in the thyroid. A special variety of epithelium covers the internal surfaces of the bladder and other parts of the urinary tract. It is waterproof and several cells thick, so that it can be greatly stretched without gaps appearing.

The outside of the body, the outer ear, mouth, throat, anus and vagina have a multilayered epithelium; the deepest cells in this divide to replace the flattened dead cells which are shed from the surface. As new cells are pushed outward they manufacture the horny substance *keratin,* and by the time they reach the top they are no more than tiny scales.

The many glands from which secretions drain through a duct to a free surface should be mentioned at this point, for in their development they are formed from ingrowths of epithelium. Such ingrowths vary in complexity from the simple coiled tubes of sweat glands to the many-branched structures found in salivary glands and the breast. They are all basically the same, however, the products of their cubical cells collecting in a central cavity and then passing along their duct.

Section of salivary gland

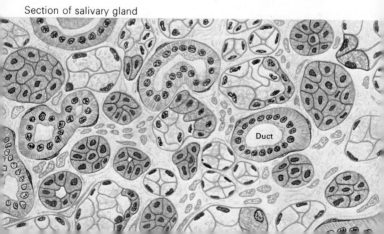

Duct

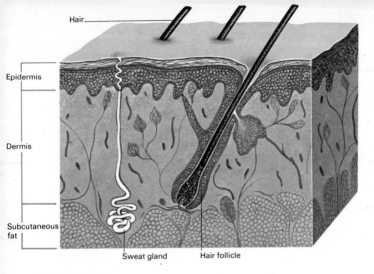

Section of skin

Skin

The skin is the largest, thickest and most complex epithelial tissue in the body and is an organ of heat regulation as well as one of protection. It has two parts – the horny outer *epidermis* and the soft inner *dermis,* in which are embedded hair follicles and sebaceous and sweat glands.

In all people except albinos, brown granules of melanin are contained in the deepest cells of the epidermis. This pigment is made in special cells (melanoblasts) lying just below the epidermis; melanoblasts are activated by ultra-violet light, producing freckles and tanning in light skins.

Hairs grow upwards from the dermis and are continually shed and replaced, new hairs forming in old follicles. Attached to the follicles are strands of smooth muscle. When they contract they pull the hairs upright (trapping air and creating a heat-insulating layer), causing gooseflesh.

Sebaceous glands open into hair follicles and lubricate the skin with their oily secretion, which is produced by the disintegration of the glandular cells. They are distinct from sweat glands, the watery secretion of which evaporates on the surface, helping to cool the body. The dermis has many blood vessels, and sweat glands especially are richly supplied.

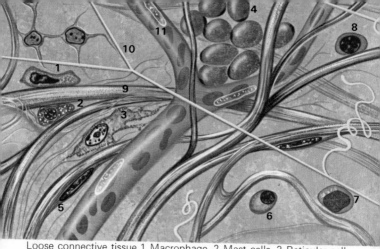

Loose connective tissue 1 Macrophage. 2 Mast cells. 3 Reticular cell. 4 Fat cells. 5 Fibroblast. 6 Plasma cell. 7 Monocyte. 8 Lymphocyte. 9 Collagen fibres. 10 Elastic fibres. 11 Blood vessel

Connective tissue

Connective tissue has relatively few cells surrounded by a mass of intercellular material and fibres. It provides a mobile supporting framework for more specialized tissues.

In the loose tissue beneath the dermis there is a viscous ground substance criss-crossed by branching elastic fibres and bundles of tough collagen fibres. Between these are a variety of cells, including spindle-shaped *fibroblasts* (which make collagen); *macrophages* (amoeba-like defensive cells, capable of *phagocytosis* – engulfing foreign or dead material); *plasma cells* and *lymphocytes* (involved in immunity); and *mast cells* (which manufacture the ground substance). Capillary blood vessels and nerves pass through connective tissue, and fat cells are also found.

Tendons are formed of strong white fibrous tissue composed of parallel bundles of collagen fibres. Yellowish elastic tissue is found in the trachea, large arteries and spinal ligaments, and is tough but springy.

Most of the fat in the body is stored in connective tissue inside special cells, the cytoplasm of which is stretched into a thin bubble. In fatty areas all other elements are displaced, resulting in a uniform appearance and a soft texture.

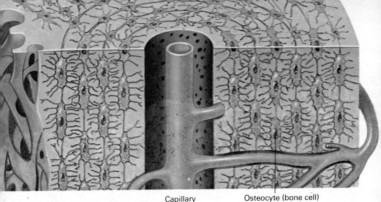

Capillary   Osteocyte (bone cell)

Structure of compact bone

The tissues of the skeleton

Two types of firm material, bone and cartilage, form the inner skeleton of the human body. In early foetal life, all 'bones' are made of cartilage; this is more flexible than bony tissue and is later replaced in all weight-bearing parts.

Most bones have a compact outer layer and a porous inner part. Some long bones are hollow with a central marrow cavity. All have a fibrous covering, the *periosteum,* containing cells which can form new bone to mend fractures. Compact bone consists of cylindrical layers surrounding intercommunicating canals, through which pass nerves and blood vessels. The bone cells are harboured in small cavities between the layers; these cavities are connected with each other by fine channels filled with cell branches. The substance between the cells is the important part of bone, for it is hard because of impregnation with crystals of minerals.

Cartilage (gristle) is semi-transparent and elastic. It is found in adults in the nose, external ear, larynx and air passages, the front part of ribs, and covering the moving surfaces of some joints. It consists of a dense feltwork of fine collagen fibres

Structure of cartilage

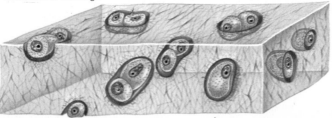

encased in a mucoid substance and containing small isolated groups of cells. In certain parts of the body, such as intervertebral discs, it is strengthened by the incorporation of thick bundles of collagen, while in other places, like the external ear, it contains elastic fibres.

The transformation of cartilage into bone begins in the foetus with the appearance of centres of bone production in each piece of cartilage. Bone-forming cells make a fibrous matrix on which calcium is deposited, and which extends outwards until virtually complete replacement has occurred. A plate of proliferating cartilage cells, however, remains near the ends of bones until adolescence; the newly-made cartilage becomes calcified, enabling the bones to lengthen.

The joints between bones are of several different types. Some allow no movement, like the fibrous junctions between skull bones. Others, like the vertebral joints with their intervening discs, permit limited bending and rotation. A third group, with a lubricated membrane and cartilaginous discs on moving surfaces, gives free movement.

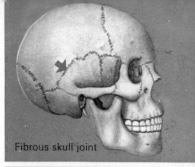
Fibrous skull joint

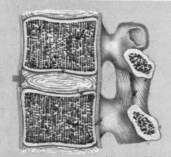
Vertebral joint

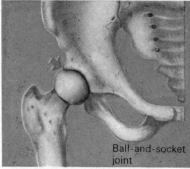
Ball-and-socket joint

Hinge joint

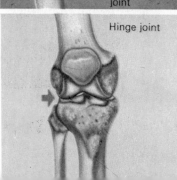

CIRCULATION

The anatomy of the thorax

The chest or thorax is enclosed below by the diaphragm and above by the clavicles, upper ribs and lower neck vertebrae. It contains and protects the heart and lungs, which are surrounded by a flexible wall composed of ribs connected by intercostal muscles and covered with soft tissues and skin.

The twelve thoracic vertebrae, behind, and the curved ribs and sternum, at the sides and in front, form a protective bony framework for the vital organs of the chest. Greater protection

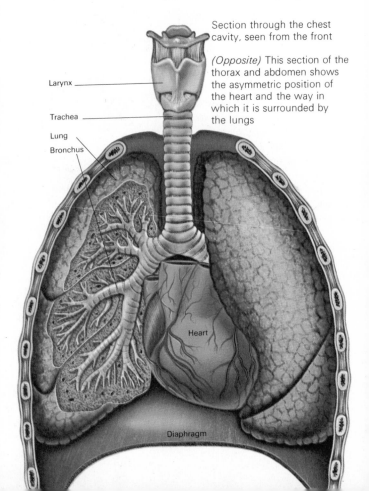

Section through the chest cavity, seen from the front

(Opposite) This section of the thorax and abdomen shows the asymmetric position of the heart and the way in which it is surrounded by the lungs

Larynx

Trachea

Lung

Bronchus

Heart

Diaphragm

would come from a solid chest cage, as is found in some invertebrates, but a rib-cage has the advantages of greater mobility and less weight.

The diaphragm is a muscular and fibrous sheet which divides the thoracic from the abdominal cavity. Its margins are attached to the inner surfaces of the lower ribs, and the contractile muscle fibres run horizontally and radially towards the centre of the chest where they become tough, inelastic and fibrous. Several important structures pass through the diaphragm. These include the oesophagus, leading to the stomach; the descending aorta, which carries arterial blood to the abdomen and the legs; and the inferior vena cava, through which blood returns from these parts to the heart. Each has its own opening in the diaphragm, and vagus nerves travel down with the oesophagus, and sympathetic nerves and lymph channels with the aorta, while the nerves of the diaphragm are close to the upper part of the inferior vena cava.

The heart sits on the front and centre of the diaphragm

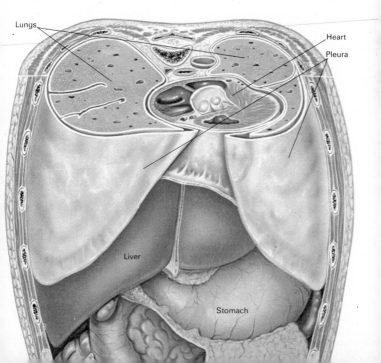

and is surrounded by the lungs, which fill the rest of the thoracic cavity. The tip of the left ventricle is in contact with the inside of the chest wall between the fourth and fifth ribs 4 inches to the left of the sternum; when the left ventricle contracts, a localized pulsation can be felt at the apex point – the 'apex beat'. When the heart is enlarged, the apex beat is displaced outwards, beyond its normal limit of 4 inches from the middle of the sternum and downwards so as to be felt below the fifth rib.

The aorta emerges from the left ventricle and rises for 2 to 3 inches before arching backwards towards the spine and descending. The arch of the aorta passes to the left of the trachea and above the left pulmonary artery.

The functions of the circulation

The heart and blood vessels form a completely closed system. Blood leaves the heart and travels via the aorta to strong, muscular, arterial vessels, which divide into smaller muscular vessels, the arterioles. These progress into tiny, thin-walled capillaries, which ramify between the cells and supply them with

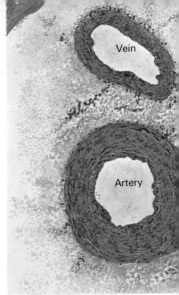

This cross-section shows the relative thickness of the walls of a vein and an artery

The arterio-venous bridge

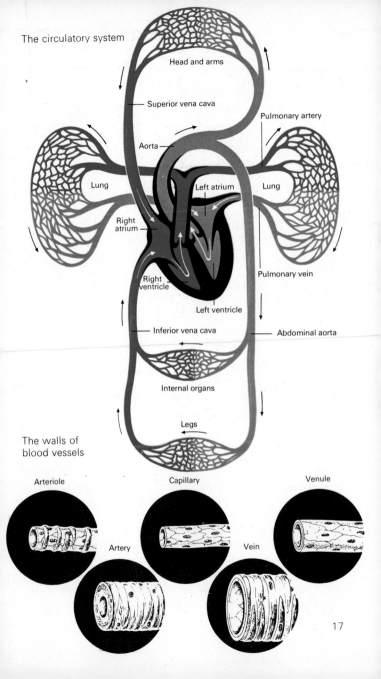

The circulatory system

Head and arms

— Superior vena cava

Pulmonary artery

Aorta —

Left atrium

Lung

Lung

Right atrium —

Right ventricle

Pulmonary vein

Left ventricle

— Inferior vena cava

— Abdominal aorta

Internal organs

Legs

The walls of blood vessels

Arteriole

Capillary

Venule

Artery

Vein

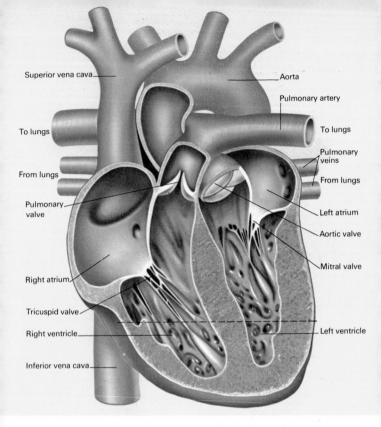

Superior vena cava
Aorta
Pulmonary artery
To lungs
To lungs
Pulmonary veins
From lungs
From lungs
Pulmonary valve
Left atrium
Aortic valve
Mitral valve
Right atrium
Tricuspid valve
Right ventricle
Left ventricle
Inferior vena cava

oxygen, glucose, hormones and other essential substances.

The function of the circulation is to deliver energy and food supplies to each cell and to remove waste products of cell metabolism like carbon dioxide. If such products were not removed, they would interfere with the normal workings of the cells and ultimately cause their death; but the circulation removes them from where they are produced and carries them to the lungs, kidneys and liver, where they are excreted or changed into less toxic substances.

The capillaries coalesce into venules, leading into larger veins which eventually drain into the right atrium of the heart by way of the two venae cavae.

The chambers and valves of the heart

The heart has four chambers, the right and left atria and the right and left ventricles. The ventricles empty blood into the pulmonary arteries and aorta, while the atria collect blood from veins. Arterial blood pressure is much higher than venous and the walls of the ventricles are correspondingly thicker than those of the atria.

The aortic and pulmonary valves are placed at the outflow of the left and right ventricles, and mitral and tricuspid valves divide off the left and right ventricles from the adjacent atria. The cusps of the valves float freely in the blood passing through them, and their position depends on the pressure difference on their two sides. When the valves close the cusps come together to form a blood-tight membrane.

Blood from the tissues enters the right atrium via the venae cavae. This blood has relatively little oxygen and a relatively high carbon dioxide content (partial pressures 40 mm and 46 mm of mercury respectively), and it passes through the tricuspid valve into the right ventricle, from where it goes into the pulmonary artery and to the lungs. The blood pressure on the right side is low (25 mm of mercury in the ventricle) and so the chambers are relatively thin-walled. In the lungs carbon dioxide is removed and the blood oxygenated. Blood returns to the left atrium via the pulmonary veins, and now has relatively more oxygen and less carbon dioxide (oxygen 100 mm, carbon dioxide 40 mm). It then flows through the mitral valve into the thick left ventricle and is pumped out through the aortic valve into the aorta, from where it is distributed throughout the body. Blood pressure in the aorta is about 120 mm of mercury.

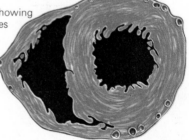

(Opposite) Section of the heart showing chambers, major vessels and valves

The left ventricle operates at higher blood pressures than the right. This section through the ventricle at dotted line on opposite page shows clearly the thicker muscle wall of the left ventricle.

The blood supply of the heart

The right and left coronary arteries supply blood to the heart muscle. They arise from the aorta just after its origin from the left ventricle. The right coronary curves around the right border of the heart, passing between atrium and ventricle, and supplies most of the back and the under-surface of the heart with blood. The left one runs in the opposite direction to supply the left side of the heart, and it gives off a large branch which descends in front between the ventricles and provides blood for this region. Disease which causes narrowing or partial blockage of these arteries is responsible for heart attacks ('coronaries').

The cardiac cycle

The cardiac cycle is the name given to the series of events which occurs in a single heart beat. During each cycle the two atria contract together and then the ventricles contract, and blood is forced from the atria into the ventricles and from these into the aorta and the pulmonary artery. This takes place 50 to 80 times a minute in normal adults.

The period in which the ventricles are contracting is called *systole* and that in which they relax *diastole*. Immediately after systole the atria begin to fill with blood (see *a* below). Tricuspid and mitral valves are closed at this time. The pressures in the ventricles fall and the pressures in the atria rise. As soon as the pressures in the atria exceed those in the ventricles the tricuspid and mitral valves float open. The ventricles, now

a Atria fill with blood.

b Contraction of atria forces more blood into ventricles.

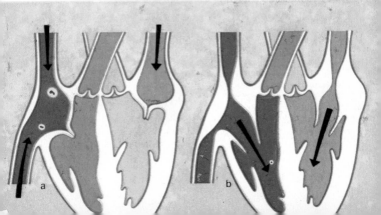

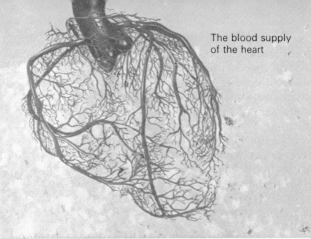

The blood supply of the heart

relaxed and dilated, fill with blood and the cardiac cycle begins with contraction of the atria (*b*), which forces more blood into the ventricles. Almost immediately the ventricular muscle contracts. The pressures in the ventricles rise very rapidly and soon exceed those in the atria, so that the mitral and tricuspid valves are forcibly closed (*c*).

Whilst these valves are closing, the ventricular pressures are greater than those in the aorta and pulmonary artery, and so the aortic and pulmonary valves are thrown open. Blood is forced smoothly through them by continued ventricular contraction, at the end of which the ventricles relax, the pressure falls, the aortic and pulmonary valves close and the cycle begins again (*d*).

c Ventricles force blood into aorta and pulmonary artery.

d Valves close

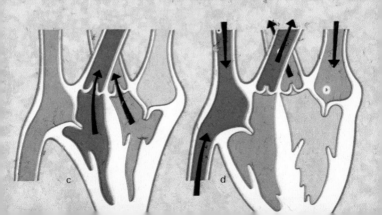

Excitation of the heart

The muscular contractions of the heart are produced by regular electrical impulses originating in a tiny area of muscle, the sino-atrial node or 'pacemaker', at the junction of the right atrium and superior vena cava. This node contracts spontaneously, and this stimulates the walls of the atria, from which the contractile impulse reaches another special area, the atrio-ventricular node. From here the impulse travels at 4 metres per second in the fibres of the 'bundle of His' – muscle fibres adapted for fast conduction – and spreads throughout the ventricles, stimulating them to contract.

The heart rate depends on the activity of the sino-atrial node, but if this is damaged the atrio-ventricular node takes over its function. This node has its own contraction rate of 35 to 45 per minute, which is unaffected by any external influences. In contrast, the sino-atrial node is controlled by the vagus nerve (which slows its rate) and the sympathetic nerves (which increase it).

The cardiac action initiated by the sino-atrial node can be recorded in the form of an *electrocardiogram* (ECG). Electrical activity of each part of the heart has its own wave-form. The P wave and the QRS waves come respectively from the contracting atria and ventricles, while the T wave coincides with relaxation of the ventricles. Disease of the heart usually produces changes in the wave pattern.

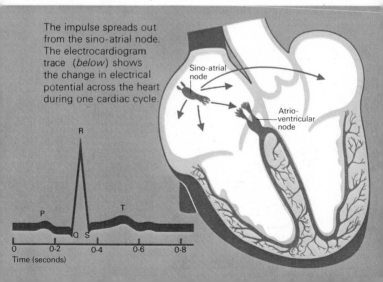

The impulse spreads out from the sino-atrial node. The electrocardiogram trace (*below*) shows the change in electrical potential across the heart during one cardiac cycle.

Sino-atrial node

Atrio-ventricular node

R

P

Q S

T

0   0·2   0·4   0·6   0·8
Time (seconds)

Heart sounds and murmurs

Passage of blood through the irregularly shaped heart cavities is turbulent, with fluid eddies which cause vibrations. These can be heard with a stethoscope as heart sounds and murmurs. In the normal heart, two sounds are produced, the second being shorter and sharper than the first. The first sound is due to closure of the mitral and tricuspid valves at ventricular systole, while the second results from closing of the aortic and pulmonary valves as the ventricles relax.

Disease of the valves roughens their surface and increases turbulence, so that vibrations may be produced while the valves are open, that is, between the heart sounds. Such vibrations constitute heart murmurs and doctors can often pinpoint the damaged valve by careful study of the murmur. Sometimes babies are born with an incomplete partition between atria and ventricles, and passage of blood through this 'hole in the heart' is another cause of murmurs. Increased turbulence, with resultant murmurs, is not always due to heart disease, but may occur in health, especially in children.

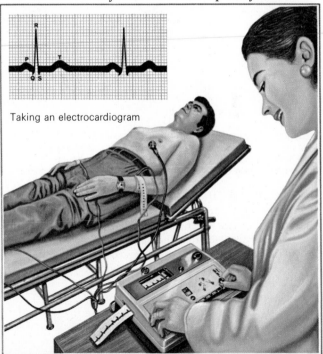

Taking an electrocardiogram

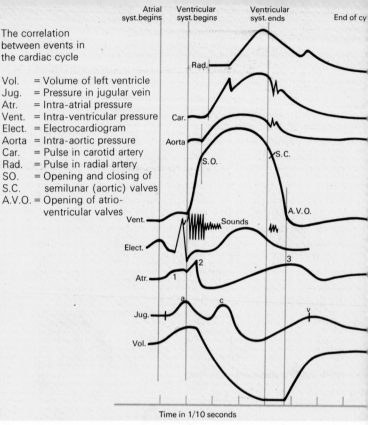

The correlation between events in the cardiac cycle

Vol. = Volume of left ventricle
Jug. = Pressure in jugular vein
Atr. = Intra-atrial pressure
Vent. = Intra-ventricular pressure
Elect. = Electrocardiogram
Aorta = Intra-aortic pressure
Car. = Pulse in carotid artery
Rad. = Pulse in radial artery
SO. = Opening and closing of
S.C.   semilunar (aortic) valves
A.V.O. = Opening of atrio-
    ventricular valves

Time in 1/10 seconds

The electrical activity corresponding to atrial and ventricular excitation immediately precedes the respective atrial and ventricular systoles. Opening and closing of the semilunar valves are critical for the aortic pressure changes and changes in ventricular volume. The pressure wave is delayed and distorted along the arterial tree.

Normal blood pressure

'Blood pressure' is the pressure of blood in the arteries and is a measure of the tension in the arterial wall produced by the blood forced through from the heart. This pressure depends on two factors – the output of the heart, and the resistance to flow provided by the smaller arteries and arterioles. In young adults blood pressure normally reaches 120 mm of mercury during systole and falls to 80 mm in diastole.

Cardiac output is the volume of blood pumped out per minute, and this will obviously increase if the heart rate goes up. Resistance to blood flow depends on the diameter of the smaller muscular vessels, which can be constricted by activity of the sympathetic nervous system, itself influenced by the vasomotor centre. This centre is in the hind-brain, and it sends out impulses which pass down to the sympathetic nerves supplying the vessels. It is regulated by nervous messages, especially those coming to the brain from sensitive pressure receptors in the wall of the aorta, and normally these receptors are sending signals to the centre all the time. Blood pressure is thus constantly monitored and adjusted to a more or less steady level. Any increased pressure in the aorta, for example from an increased cardiac output, is signalled to the vasomotor centre, which suppresses sympathetic activity – causing relaxation of small vessels and a fall in resistance to flow. Conversely, a fall in aortic pressure produces increased activity in the centre and stimulation of sympathetic nerves; the arterioles contract and their resistance increases, helping to maintain the blood pressure at its usual level.

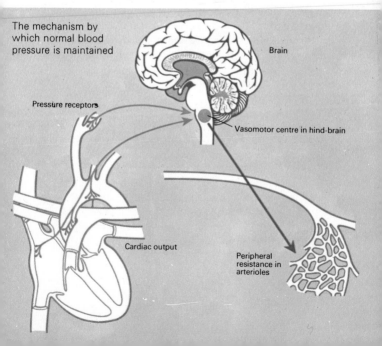

The mechanism by which normal blood pressure is maintained

Brain

Pressure receptors

Vasomotor centre in hind-brain

Cardiac output

Peripheral resistance in arterioles

High blood pressure (hypertension)

High blood pressure (hypertension) results from increased resistance to flow. This may be because vessels are narrowed or because the blood is more viscous than normal. Increased cardiac output does not cause sustained hypertension, but can produce a transient rise.

Emotional stresses of all kinds are reflected in the blood pressure. Excitement, agitation or annoyance stimulates sympathetic nervous activity, which accelerates the heart and contracts arteriolar muscle, both of which effects cause hypertension. A heated argument can raise blood pressure to as much as 240 mm (systolic)/130 mm (diastolic) – normal values are 120/80.

Another important factor is the influence of certain hormones, especially *noradrenaline*. As described later in the book, this substance is produced by the medulla of the adrenal gland and it causes contraction of the muscular walls of blood vessels. If too much is in the circulation, this contraction results in a prolonged rise of blood pressure. This may occur with a tumor of the adrenal medulla, and if this is removed the blood pressure may return to mormal.

Kidney disease plays an important part in producing high blood pressure, for the damaged kidney secretes a substance called *renin* which is converted in the blood to another substance which is very active in producing arteriolar contrac-

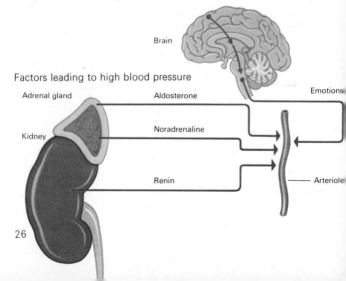

Factors leading to high blood pressure

Brain

Adrenal gland

Aldosterone

Emotions

Kidney

Noradrenaline

Renin

Arteriole

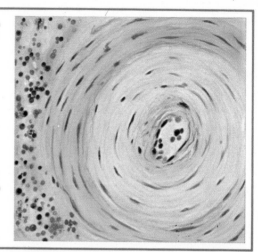

This transverse section of a small artery in the kidney of a case of malignant hypertension shows the thickening of the vessel walls.

(*Below*) Severe hypertension can damage the retina of the eye. (*Left*) Normal appearance and (*right*) severe hypertension, with haemorrhage of blood vessels and flecks of hard exudate.

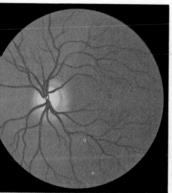
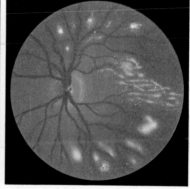

tion. Renin is secreted in a wide variety of kidney diseases, many of which cannot be cured. One, however, which can be treated is narrowing (stenosis) of the renal artery; here large quantities of renin are produced and the hypertension may be severe, but surgical correction of the narrowing often results in a cure.

The blood becomes more viscous when the proportion of red blood cells increases. This occurs in *polycythaemia*. A greater force than normal is needed to push the thick blood through the vessels, and this is shown as an increased arterial pressure.

Finally, overloading of the circulatory system with blood may cause hypertension, for the vessels become filled beyond their normal capacity. The volume of blood may increase as a result of salt and water retention by the kidneys, as occurs when tumours of the adrenal cortex secrete aldosterone.

The walls of the muscular arteries and arterioles respond to an increased pressure by thickening. This in turn reduces the vessel diameter, cutting down the flow of blood. The brain, the kidney and the heart may be seriously or fatally damaged by the decreased blood supply. A greater force is needed for the heart to impel blood through the raised resistance of the circulatory system if adequate flow is to be maintained. This of itself implies an even greater arterial blood pressure and a consequent worsening of the situation.

Progressive changes in the circulatory system occur with age. The gradual silting up and hardening of arterial and arteriolar vessels lays an ever increasing strain on the ageing heart.

Shock

Shock, in the medical sense, is a condition characterized by prolonged low blood pressure, rapid pulse, sweating, cold hands and feet, and eventually loss of consciousness. These features are a result of the body's tendency, when the total

The reaction of the body to rapid bleeding

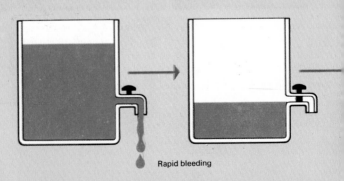

Rapid bleeding

volume of blood is too small, to divert blood to the organs which above all others need oxygen and glucose – the brain and kidneys.

Under normal conditions about 15 per cent of the output of the heart goes directly to the brain and 25 per cent to the kidneys. The supply to muscles accounts for about 20 per cent of the cardiac output; it should be remembered that muscles make up two-fifths of the total body weight, so on a simple weight-for-weight basis, brain and kidney blood flow is much greater.

When the volume of circulating blood is low, as occurs after severe rapid bleeding from gross injuries, or after severe water loss (as in acute diarrhoea), the circulation reacts by selectively shunting blood to the brain and kidneys. In order to maintain blood pressure and flow the small vessels in muscle and skin contact – in response to stimuli from the vasomotor centre – and thus a shocked patient appears cold and white. The adrenal glands are stimulated, and pour out adrenaline and noradrenaline, while sympathetic nerves are overactive. An increased heart rate and sweating are two of the results.

If bleeding or fluid loss continues, the body is unable to cope, and despite the increased heart rate and severe arteriolar constriction the blood pressure cannot be maintained. It falls, the blood supply to the brain becomes inadequate and consciousness is lost. Unless blood volume is rapidly returned to normal, permanent brain damage and death will ensue.

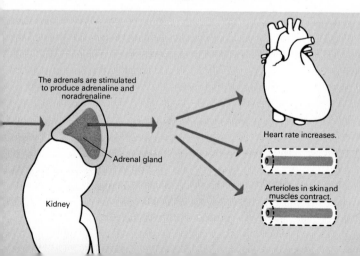

The adrenals are stimulated to produce adrenaline and noradrenaline.

Adrenal gland

Kidney

Heart rate increases.

Arterioles in skin and muscles contract.

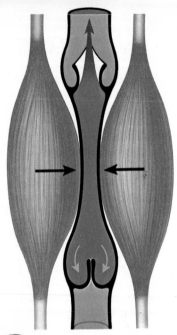

Fainting

Fainting is loss of consciousness in the presence of an adequate circulating blood volume, and is due to a short-lived decrease in blood flow to the brain.

The blood in the veins is helped on its way back to the heart by the contractions of the limb muscles, especially those in the legs, which squeeze the veins and act as a pump. Venous pressure depends on gravity, and this muscle pump prevents blood accumulating in the legs in the upright position. If someone stands still for a long time the leg muscles are not very active and so blood tends to pool there. The heart can only circulate the blood it receives, and the fall in the volume of blood returning to the heart may be great enough to decrease the cardiac output to the point where the blood flow to the brain suffers and unconsciousness follows. This is why such fit young people as guardsmen faint on parade.

Fainting may also occur when the heart rhythm is abnormal, for such abnormality may also lower the cardiac output. 'Vaso-vagal attacks' are faints especially common in young girls: here

(Above) The muscle pump. Contraction of the muscles compresses the vein, forcing the blood through the upper valve, while the lower valve prevents backflow.

(Left) When a person is in a standing position for a long time, blood pools in the legs. *(Below)* After fainting, the return of blood to the heart from the veins is no longer impeded.

it is thought that emotional upset causes increased activity of the vagus nerves, which slow the heart and reduce its output. In some heart diseases the conducting system is damaged and unable to function, so that the ventricles beat at their own rate of 35 to 45 per minute. The result is 'heart block', and this slow rate may be inadequate for brain blood supply, so that attacks of unconsciousness frequently occur. These can sometimes be corrected by inserting an artificial pacemaker, an electrical device which stimulates the heart to contract rhythmically at a normal rate.

The result of unconsciousness is falling to the ground. The body becomes horizontal, and in this position the effect of gravity impeding venous return and pooling blood in the legs disappears, and the flow of blood to the heart from the veins is improved. Cardiac output increases, the blood supply to the brain returns to normal and consciousness is restored.

There are a variety of types of pacemaker. Shown here is a P wave pacemaker, which takes over when the sino-atrial node fails to evoke the natural QRS response. It amplifies the P wave, delays it for an appropriate interval and generates a pulse which evokes the QRS wave.

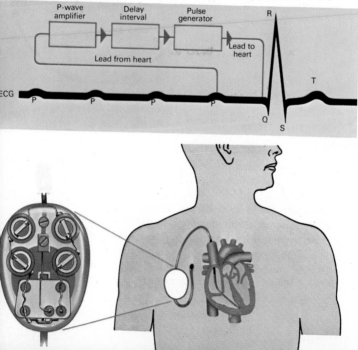

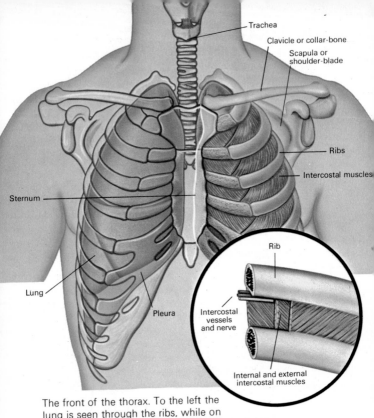

The front of the thorax. To the left the lung is seen through the ribs, while on the right the ribs and intercostal muscles are shown.

RESPIRATION

The events in breathing

Breathing supplies the body with oxygen and removes carbon dioxide from it. Energy production in the cells needs oxygen; carbon dioxide is a by-product of energy release, and if it is allowed to build up it will poison the cells.

Ventilation of the lungs is brought about by contraction of the diaphragm and to a lesser extent the intercostal muscles. When its fibres contract, the diaphragm moves downwards. During quiet breathing the range of movement is a half to one inch. When oxygen requirements are greater, breathing becomes deeper and faster and the movement of the diaphragm increases three to fourfold. The intercostal muscles are attached to the upper and lower margins of the ribs. They are in two layers, with fibres running in different directions, so arranged that when they contract they pull the ribs closer together. Because of the way in which the ribs are jointed to the vertebrae this conduction rotates each rib upwards and outwards, like a bucket handle, and a similar movement of all the ribs produces expansion of the chest. Although most of the work of breathing is done by the diaphragm, the intercostals can take over completely if the diaphragm becomes paralysed.

At the start of a breath, just before breathing in, the pressure inside the lungs is the same as the atmospheric pressure (760 mm of mercury). Then the intercostal muscles contract, moving the rib cage upwards and outwards, and at the same time the diaphragm contracts and descends. As a result the size of the chest cavity is increased and the intrathoracic pressure falls by 2 to 3 mm of mercury. At this stage the atmospheric pressure is more

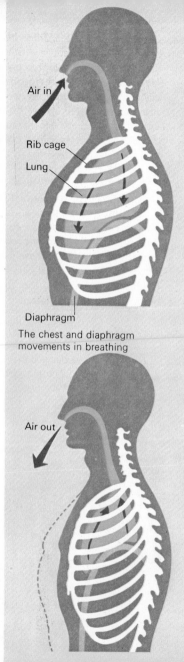

The chest and diaphragm movements in breathing

than the pressure inside the chest; since the gases in the lungs are in continuity with those outside, air rushes into the lungs to equalize the pressure difference.

At the end of each inspiration the diaphragm and intercostals relax. The diaphragm and chest wall return to their previous positions, and this fall of chest volume, together with the elasticity of the lungs, forces used air back into the atmosphere. In quiet breathing the pressure within the lungs varies from —3mm of mercury (relative to atmospheric pressure) in inspiration to + 3 mm during expiration.

Air, which contains about 20 per cent of oxygen and 79 per cent of nitrogen, is breathed in through the mouth and nose, passes into the larynx, which is guarded by the epiglottis, and enters the major airway, the *trachea*. This is a firm tube held open at all times by crescents of tough cartilage arranged one above the other. These rings do not meet at the back, and the trachea is composed behind of fibrous tissue. In section the trachea is D-shaped, the posterior part – in front of the oesophagus – being flattened.

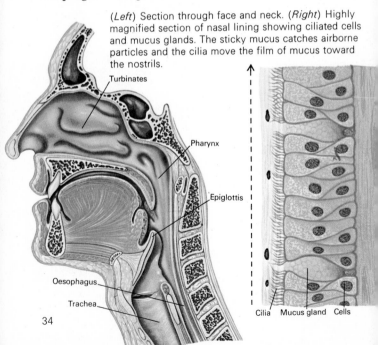

(*Left*) Section through face and neck. (*Right*) Highly magnified section of nasal lining showing ciliated cells and mucus glands. The sticky mucus catches airborne particles and the cilia move the film of mucus toward the nostrils.

Turbinates

Pharynx

Epiglottis

Oesophagus

Trachea

Cilia   Mucus gland   Cells

In the chest the trachea divides into the two main *bronchi,* right and left, which are also cartilaginous. These divide further into branches supplying the lobes of the lungs. Thus on each side there are upper and lower lobe bronchi. The right lower lobe bronchus gives off a branch to the middle lobe; on the left the counterpart to this is the lingular bronchus, which comes from the upper lobe bronchus. These air channels branch again and again, and their smallest offshoots, which are only a few thousandths of a millimetre in diameter, open finally into blind-ended cavities called *alveoli.*

The lungs are encased by the *pleura,* a very thin glistening membrane which enables them to move freely and without friction inside the chest wall. The pleura has two layers: the outer lines the rib cage and intercostal muscles and the inner covers the pulmonary lobes, the two layers joining where the bronchi and blood vessels enter the lung. Normally they are closely pressed against each other, but there is always a potential space, the pressure in which is sub-atmospheric. This space can fill with air or fluid if the pleura becomes inflamed or injured, while pleurisy, accompanied by chest pain on breathing, results from an inflammatory roughening of these normally smooth surfaces.

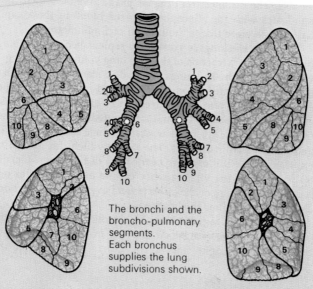

The bronchi and the broncho-pulmonary segments. Each bronchus supplies the lung subdivisions shown.

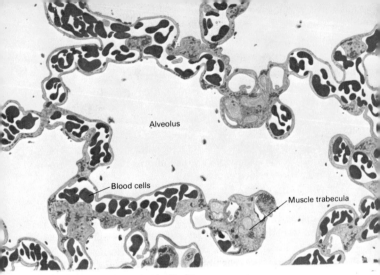

Microscopic section of alveoli of lung

The movement of gases into the lungs

The alveoli, the blind openings of the terminal bronchioles, are lined by a single thin layer of cells on a fine basement membrane, attached to which are the walls of capillary blood vessels. The distance from the air in the alveoli, which communicates with the atmosphere, to the blood in these capillaries, is no more than a thousandth of a millimetre.

The air in the lungs is constantly replenished in breathing, but in fact only a fraction of the total lung volume is exchanged with each breath. The lungs of an adult man contain about 6 litres of air. About half a litre is breathed in and out in quiet respiration, but up to 4 or 5 litres can be exhaled with a great effort after as deep an inspiration as possible. This is the so-called vital capacity; it decreases with age and decrepitude and is increased in athletes. The rate at which this air can be expelled is a useful measure of the state of the lungs. Usually 80 per cent of the vital capacity can be expired within one second, but disease of the airways (and even smoking just one cigarette) can reduce this markedly.

Movement of fresh air into the alveoli is essential for the oxygenation of venous blood arriving in the alveolar capillaries from the right side of the heart. The blood in the

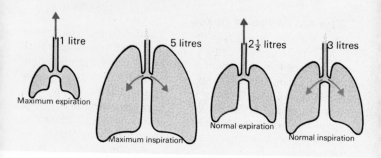

Lung capacities in deep breathing and normal breathing

capillaries arrives with an oxygen tension (partial pressure) of 40 mm of mercury and an oxygen saturation of 70 per cent, as well as a carbon dioxide tension of 46 mm. The fresh alveolar air has an oxygen tension of 100 mm of mercury and virtually no carbon dioxide. With these great pressure differences rapid equilibration takes place, carbon dioxide going into the alveoli and oxygen into the capillary blood. Fully oxygenated blood (tension 100 mm, saturation 100 per cent) with a carbon dioxide tension of 40 mm leaves the lungs for the left side of the heart and the aorta.

(*Left*) The exchange of gases through the alveolar walls.
(*Right*) A group of alveoli showing blood supply

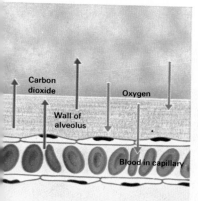

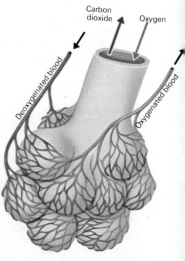

The exchange of gases

Haemoglobin, which fills the red blood cells and colours the blood, is a chemical compound which readily combines with oxygen and readily releases it. The process of combination, which occurs in the lungs when blood passes through the alveolar capillaries, turns the purplish *reduced haemoglobin* into bright red *oxyhaemoglobin*. The reverse process occurs in active tissues of the body and accounts for the difference in colour between arterial and venous blood. Haemoglobin has an affinity for certain other substances. Of these the most important is carbon monoxide, found in coal gas and car exhaust fumes, which becomes attached very firmly forming crimson *carboxyhaemoglobin* – this prevents the carriage of oxygen in the blood and can cause death through oxygen lack.

The degree of oxygen saturation of haemoglobin depends on the amount of the gas in the plasma surrounding the red cells. By exposing blood samples to different concentrations of oxygen and analysing them, the way in which haemoglobin behaves can be investigated.

The oxygen dissociation curve represents the behaviour of haemoglobin in the body. The S shape of the graph illustrates several important facts. In health the tension of oxygen in the alveoli is 100 mm of mercury and this produces a tension of 100 mm in the plasma in the alveolar capillaries. The haemoglobin in the red cells in these capillaries will thus be 100 per cent saturated. In the fluids surrounding the cells of the body

Normal solar spectrum (1) absorption spectra of oxyhaemoglobin (2) reduced haemoglobin (3) and carboxyhaemoglobin (4).

the oxygen tension is the same as that inside the cells – 35 to 40 mm. Fully saturated haemoglobin is unable to hold on to its oxygen, which is released and becomes freely available for use by the cells. Venous blood, coming from the capillaries surrounding the cells, is thus unsaturated (oxygen tension 40 mm, oxygen saturation 70 per cent).

Cells can survive if the blood supplying them contains only 70 per cent of the normal amount of oxygen. However, if for any reason the oxygen tension falls below 40 mm of mercury, the oxygen saturation of the blood falls precipitously; oxygen supply to the cells decreases to zero and they die.

The graph plots percentage of haemoglobin combined with oxygen against *dissolved* oxygen concentration in the presence of different levels of carbon dioxide. The blue curve is without carbon dioxide, the black curve corresponds to the carbon dioxide level of arterial blood and the red curve corresponds to that of venous blood. The arrows indicate shifts in *combined* oxygen content which result from changing carbon dioxide concentration *at the same dissolved oxygen concentration*. In *a* the oxygen content rises sharply as carbon dioxide is removed and in *b* the content falls as carbon dioxide is added.

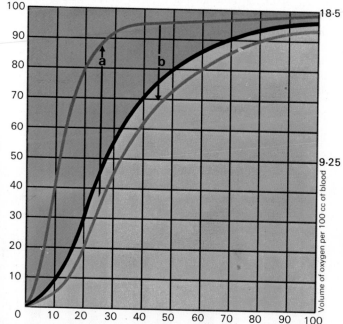

Percentage of saturation of blood with oxygen

Volume of oxygen per 100 cc of blood

Oxygen pressure in mm of mercury

Oxygen lack and excess

A fall in the body's oxygen supply can come about in two ways: the transporting system in the blood may be normal and available oxygen insufficient, or the transporting mechanism may be deficient while adequate oxygen is available. The first, which occurs more often, results from decreased oxygenation of the alveoli in the presence of normal quantities of haemoglobin. In the second, there is not enough normal haemoglobin in the blood to combine with the adequate amount of oxygen in the lungs.

Decreased alveolar ventilation often occurs in lung disease. In chronic bronchitis the walls of many of the alveoli are destroyed, resulting in an inefficient distribution of air within the lungs. In addition there are often small areas of infection, and these parts of the lung cannot be filled with air. The effect of these changes is to decrease the total surface area over which the transfer of oxygen can take place, and in consequence the blood which flows through the lungs has a smaller oxygen content than normal. Decreased oxygenation is also a feature of those lung diseases in which there is a thickening of the alveolar and capillary wall; here oxygen transfer may be markedly impaired by what is, in effect, a mechanical barrier. It is something which is found too in patients with a 'hole in the heart' – more accurately a hole in one of the partitions separating the chambers of the heart. If such a defect is present, deoxygenated blood may flow directly from

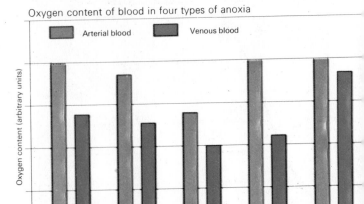

Oxygen content of blood in four types of anoxia

the right atrium or ventricle to the left side of the heart without passing through the lungs, and the arteries contain blood of low oxygen saturation. If, as a result of heart or lung disease, there is more than about 30 per cent desaturation (or about 5 grams of reduced haemoglobin per 100 mls of blood), the lips and fingers appear bluish – *cyanosis*.

Another cause of poor oxygenation is decreased oxygen tension at high altitudes. Symptoms experienced by mountaineers – fatiguability, lightheadedness, breathlessness – diminish after several days because the body can compensate for the low alveolar oxygen content by increasing the production of red cells and so making available more haemoglobin.

Decreased blood oxygen-carrying power occurs in anaemia and when haemoglobin is unusable, as in carbon monoxide poisoning.

Under certain conditions – especially in high concentration and with increased pressure – oxygen becomes toxic. It may inactivate some enzymes and produce cell damage, particularly in the nervous system.

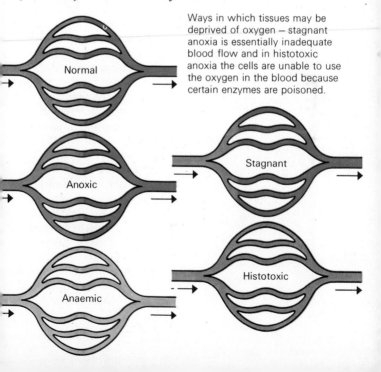

Ways in which tissues may be deprived of oxygen – stagnant anoxia is essentially inadequate blood flow and in histotoxic anoxia the cells are unable to use the oxygen in the blood because certain enzymes are poisoned.

Receptors sensitive to blood pressure and to the chemical composition of the blood are situated in the large arteries. Impulses from these sites reflexly affect circulation and respiration.

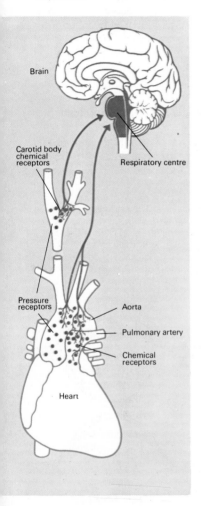

Brain

Carotid body chemical receptors

Respiratory centre

Pressure receptors

Aorta

Pulmonary artery

Chemical receptors

Heart

The regulation of breathing

Breathing depends on the activity of a group of specialized nerve cells in the medulla. These cells comprise the *respiratory centre*. Impulses from them pass down into the spinal cord and from there into the phrenic and intercostal nerves, causing contraction of the diaphragm and intercostal muscles and producing an inspiration. Impulses are sent off rhythmically, giving rise to the normal regular pattern of respiration. This rhythmicity is influenced by nerve information sent back to the medulla from the respiratory muscles and from the lungs, depending on whether they are contracted, relaxed or stretched, but it is partly an inherent property of the centre.

The respiratory centre is under many other influences. Obviously it can be affected by voluntary activity, for breath can be held until one is severely lacking in oxygen; it is however impossible to commit suicide by holding the breath and clearly other involuntary factors dominate respiration. The chemical state of the blood, in interaction with neural influences ultimately determines ventilation. The centre is sensitive to change in blood oxygenation, carbon dioxide content and acidity, not only in blood that passes through it but also in blood in the carotid arteries, changes here being signalled to the brain by sensitive receptor nerves. An increased level of blood carbon dioxide stimulates the activity of the respiratory centre both directly and indirectly, and a rise in acidity has the same effect. The effect of anoxia is different. Lowered oxygen content in arterial blood has an intense stimulatory influence on the carotid chemoreceptors and thus causes increased respiration; if for any reason the receptors are damaged the direct action of anoxia on the brain (depression) is seen and respirations are diminished.

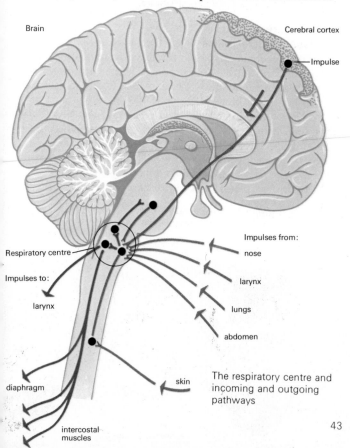

Brain

Cerebral cortex

Impulse

Impulses from:

nose

Respiratory centre

larynx

Impulses to:

lungs

larynx

abdomen

skin

The respiratory centre and incoming and outgoing pathways

diaphragm

intercostal muscles

43

The tongue of the unconscious subject *(right)* may fall back and block the airway. To clear the airway *(left)* the head should be extended and the lower jaw forced forward.

Artificial respiration

Respiratory failure occurs when the lungs are incapable of supplying adequate oxygen to the blood or removing excess carbon dioxide from it. It exists when – with a normal haemoglobin content and a normal heart – the tension of oxygen in arterial blood falls below 100 mm of mercury and the carbon dioxide tension rises above 45 mm. Artificial respiration provides a means of treating patients whose own breathing has become inadequate for their metabolic demands. It may be needed if the airways are blocked, or in cases with severe weakness of the respiratory muscles, or with depression of the brainstem respiratory centre. There are several methods,

(Left) Mouth-to-mouth resuscitation. *(Right)* External cadiac compression. Striking the heart smartly may stimulate it into beating. If there is no response, continue with mouth-to-mouth resuscitation.

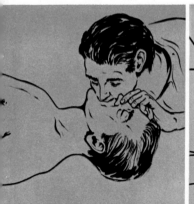
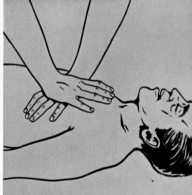

which can be used for first aid after drowning or other accidents as well as in hospital.

Mouth-to-mouth respiration, or the kiss of life, is simple and effective. The operator breathes in deeply, and then, while holding the patient's nose, he puts his mouth over the patient's mouth and expires fully. He then allows the patient to exhale. This is repeated twenty times per minute.

With the *Holger-Nielsen* technique, the patient is placed flat on his front. The chest is then compressed with two hands to the count of four. This expels air – and water – from the lungs, and is followed by relaxation, while counting from five to eight, to produce an inflow of air.

The Holger–Nielsen method. After making sure the airway is free, the operator places his hands on the patient's back (*left*) just below the shoulder blades and exerts steady pressure to the count of four. The operator releases pressure (*right*) by rocking back and lifting the patient's arms while counting from five to eight.

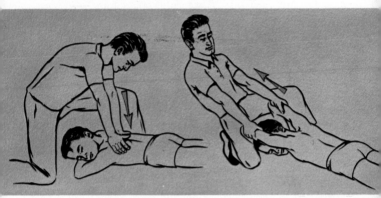

Paralysis of the diaphragm and intercostal muscles, which may occur in severe poliomyelitis, necessitates some form of mechanical aid for breathing. In such circumstances *intermittent positive pressure respiration* is the method now generally used. It consists basically of a bellows opened and closed regularly by an electrically driven motor and pumping air into the lungs via a tube in the trachea, usually inserted through an opening or tracheostomy. With modern equipment, normal breathing can be effectively simulated.

Temperature regulation

Health depends basically on the normal functioning of the biochemical processes of the body. The rate of all chemical reactions is influenced by heat, but biochemical reactions which involve enzymes are especially sensitive. Enzymes have an optimum working temperature – in humans this is $37°C$; a fall of even a few degrees slows them severely, while a rise of five or six degrees begins to destroy them. The body has thermostatic mechanisms which ensure a constant temperature and therefore constancy of chemical reaction rates.

Temperature control is achieved through a balance of heat gain and heat loss. Heat gain is partly from within and partly from outside. Cellular activity involves heat production of the order of 1700 calories per day in an average sized man. Muscular work increases this basal level of heat production tenfold or more over short spells, shivering – with repeated rapid contractions of many muscle groups – being one of the most calorific activities in normal life. Heat can also be gained from the environment, either from direct solar radiation or from hot air.

In cold climates losing heat usually presents no difficulty, but with a high atmospheric temperature or after strenuous work, body heat must be rapidly and efficiently dissipated. This is achieved by sweating, which is the body's most important means of losing heat. Up to 12 litres of sweat can be secreted in 24 hours. In evaporating from the skin, each litre uses up 580 calories, and the quantity of heat that can be eliminated this way far exceeds what can be removed in the breath and by radiation, convection and conduction into cooler surroundings. Obviously much salt and water as well as heat is lost in profuse sweating, and must be replaced if heat-stroke is to be avoided.

Body temperature is regulated by the hypothalamus in the brain, which contains nerve cells sensitive to thermal changes in their blood supply. A fall in blood temperature puts into action the involuntary nerve impulses that produce vaso-constriction and shivering, while a rise in temperature provokes heat loss by stimulating vasodilatation and sweating. Injury to this part of the brain will impair the ability of the body to keep its temperature constant.

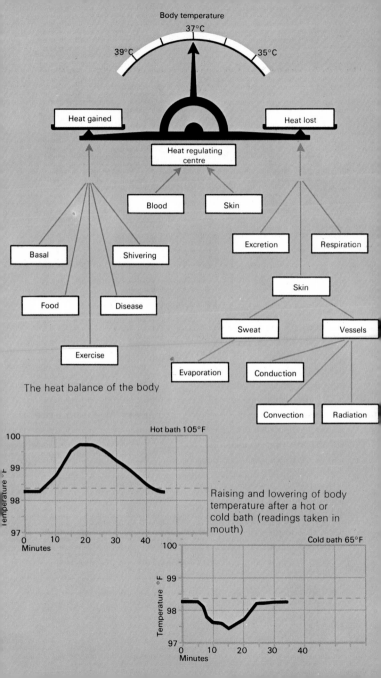

Body temperature

37°C

39°C   35°C

Heat gained

Heat lost

Heat regulating centre

Blood   Skin

Basal

Shivering

Food

Disease

Exercise

Excretion   Respiration

Skin

Sweat   Vessels

Evaporation   Conduction

Convection   Radiation

The heat balance of the body

Hot bath 105°F

Raising and lowering of body temperature after a hot or cold bath (readings taken in mouth)

Cold bath 65°F

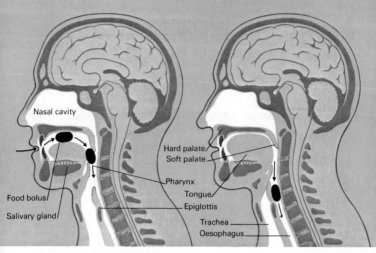

Nasal cavity

Hard palate
Soft palate
Pharynx
Tongue
Epiglottis
Trachea
Oesophagus

Food bolus
Salivary gland

The progression of a lump of food into the oesophagus

DIGESTION AND METABOLISM

Chewing and swallowing—the salivary glands

The digestion of food begins in the mouth, the teeth mechanically breaking up what is eaten, the saliva providing lubrication and the tongue mixing and moulding the food in preparation for swallowing.

The incisor teeth tear off small pieces, which are ground up by the premolars and molars. Saliva enters the mouth by way of ducts from three sets of glands, the parotid, sublingual and submandibular glands. As well as mucin (which lubricates) saliva contains the enzyme *ptyalin,* which starts off the chemical breakdown of starchy foods.

In the mouth the softened, moist ball of food is shaped by the tongue and is then passed backwards into the throat. Swallowing now begins, the sensitive nerve endings in the back of the mouth and pharynx setting into action a chain of complicated movements. First the soft palate rises, blocking off the nose. The vocal cords close and the larynx rises. The pharynx, narrowed by this movement, then opens up; the food mass passes backwards over the epiglottis, which guards the larynx, and into the oesophagus, which widens to catch it.

The oesophagus

The oesophagus is a muscular tube 10 inches long leading down from the pharynx through the chest and diaphragm to the stomach. Its wall has two muscle layers, an outer one with fibres running lengthwise and an inner one with circular fibres, and it is lined in its upper two thirds with squamous epithelium, like the mouth, and in its lower part with glandular mucus-forming cells.

In its resting state it is shut by the cricopharyngeus muscle above and the cardiac sphincter below. It is relaxed, often containing a little air and secretion but nothing else. When food is swallowed and enters it from above, *peristalsis* begins. This is the name given to the slow, automatic contractions that occur along the whole length of the intestines, from oesophagus to rectum, propelling their contents onward. In the case of the oesophagus two to three inches of its wall contracts and the wave of contraction then passes on towards the stomach at a rate of an inch or so per second. Peristalsis is very effective – so effective, in fact, that fluids will get to the stomach even in an upside-down position.

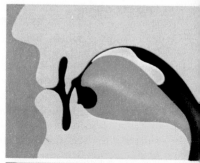

(Above) The tongue mixes and moulds food into an easily swallowed mass.

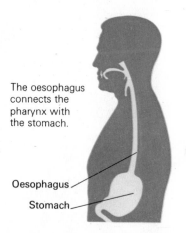

The oesophagus connects the pharynx with the stomach.

Oesophagus

Stomach

49

The stomach

The muscular bag of the stomach retains food for several hours, during which time digestion of protein takes place. There are three layers in the muscle coat, the middle set of fibres running obliquely, while at the exit of the stomach, the pylorus, the circular fibres are thickened to form a tight ring.

The lining of the stomach contains cells which secrete the enzyme pepsin, hydrochloric acid and mucus. In the presence of acid, pepsin breaks down proteins to polypeptides; mucus lines the stomach and prevents it digesting itself. Acid secretion is affected by many factors, and when disturbed peptic ulceration may result.

Smooth muscle

The muscle of the intestines (as well as certain other structures, notably blood vessels and the uterus) is called smooth muscle because under the microscope, unlike the limb muscles, it has no fine cross markings. It contracts involuntarily or auto-

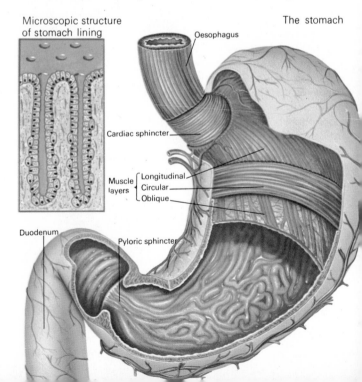

Microscopic structure of stomach lining

The stomach

Oesophagus

Cardiac sphincter

Muscle layers { Longitudinal / Circular / Oblique }

Duodenum

Pyloric sphincter

matically, in contrast to the limb muscles which respond to an effort of will, and it differs further in its relative slowness of action and in its nerve control.

Peristaltic movement is set off by distension of the hollow intestines. This affects sensory nerve cells in Meissner's plexus in the lining, and these stimulate Auerbach's plexus which lies in the muscle coats and releases acetylcholine, producing muscle contraction. Activity of vagus and sympathetic nerves can increase or inhibit peristalsis, but on the whole this is independent of external nervous influences.

In vomiting, the normal peristaltic motions of the oesophagus and stomach are abolished. In response to gastric irritation or overfilling, or stimulation of the throat or balance organs (as in travel sickness) or any sensitive part of the body, the pylorus contracts while the stomach relaxes and all the respiratory muscles go into spasm. The increased pressure in the abdomen compresses the stomach against the diaphragm, expelling the contents. Expulsion may be helped by a reversed wave of peristalsis in the oesophagus.

(*Above*) Peristalsis. (*Below*) Gastric juices, secreted in response to a signal from the brain, breakdown the food.

Contraction

Food

Canal

Gastric juices

Food

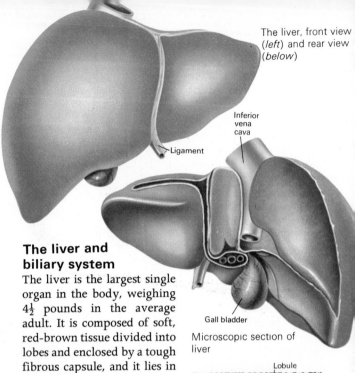

The liver, front view (*left*) and rear view (*below*)

Inferior vena cava

Ligament

Gall bladder

The liver and biliary system

The liver is the largest single organ in the body, weighing $4\frac{1}{2}$ pounds in the average adult. It is composed of soft, red-brown tissue divided into lobes and enclosed by a tough fibrous capsule, and it lies in the upper abdomen on the right side, beneath and loosely attached to the diaphragm. One remarkable feature is its double blood supply; besides receiving fresh arterial blood from the hepatic artery, which arises from the aorta, it is also fed with blood from the portal vein, carrying the products of digestion from the intestines. Its spent blood drains into the inferior vena cava via the hepatic vein.

Bile produced by the liver cells passes into the hepatic

Microscopic section of liver

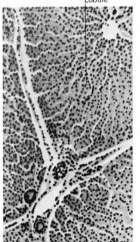

Lobule

ducts which join on the under surface of the liver into the common bile duct. This continues and meets the duct of the pancreas in a slight expansion which opens into the duodenum. The gall bladder is a blind pouch 3 or 4 inches long attached to the lower surface of the liver. It is connected to the common bile duct and receives, stores and concentrates bile.

Under the microscope the liver is seen to consist of millions of lobules in which uniform, many-sided cells are arranged in interconnecting sheets. Between the lobules run branches of the hepatic artery, portal vein and bile ducts, all closely bound together. Arterial and portal blood passes inwards into the lobules in channels called sinusoids, which are lined by the liver cells themselves, and then drains into the central vein of the lobule, which communicates with the hepatic vein. In contrast bile passes outwards in fine channels between the cells and enters the bile duct branch that is running in close proximity to arterial and portal branches. These groups of three channels, together with the adjacent parts of three lobules supplied by them, are the functional units of the liver just as the alveoli are the units of the lungs.

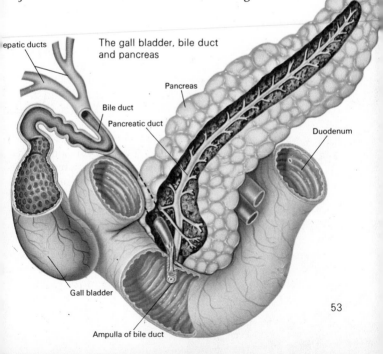

The gall bladder, bile duct and pancreas

epatic ducts

Pancreas

Bile duct

Pancreatic duct

Duodenum

Gall bladder

Ampulla of bile duct

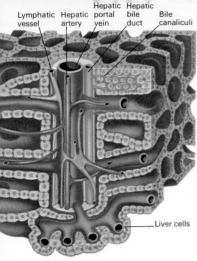

Lymphatic vessel | Hepatic artery | Hepatic portal vein | Hepatic bile duct | Bile canaliculi

Liver cells

(*Above*) Structure of the liver

(*Opposite left*) This shows the way in which the blood vessels branch within the liver and rejoin again. (*Right*) Fine structure of the liver

(*Below*) Shapes of separated liver cells

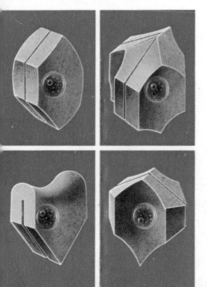

The functions of the liver

As well as being the largest the liver is also the body's most versatile organ, having an extraordinarily wide range of functions. These include production, destruction and storage, and its cells contain enzymes for many chemical processes as well as vital stores of essential materials.

The substances stored by the liver include carbohydrates, fats and proteins, and the metabolism of each of these three groups of food material takes place there. *Glycogen*, which is a condensed, readily available derivative of glucose, is deposited as granules inside the liver cells and is broken down and its products released into the blood to supply energy needs. Fat is split up in the liver and also manufactured and laid down in the cells. Many nitrogenous waste products are rendered harmless in the liver by combination with other substances; urea, the chief nitrogen-containing product of protein metabolism, is formed there from ammonia (which is toxic to the body) and carbon dioxide. The liver cells also synthesize proteins. They make the albumin and part of the globulin in the blood plasma, and the fibrinogen and prothrombin essential for blood clotting.

Up to 2 pints of bile are secreted daily. It is a viscous, dark green, alkaline fluid coloured by bilirubin, a breakdown product of haemoglobin, and containing quantities of cholesterol and bile salts. These salts have a detergent and emulsifying action, for their molecules have chemical groupings soluble both in watery and in fatty substances. This explains the ability of bile to enhance the absorption of fats from the intestines. When bile secretion is impaired, fat cannot be absorbed properly and is instead excreted in the faeces (steatorrhoea), and in addition the fat-soluble vitamins A, D and K are lost to the body.

Storage of certain vitamins is another function. Vitamin A is present in large quantities in liver fat, and vitamin D is also found (hence the value of cod-liver oil). The liver contains vitamin B_{12}, enabling the maturation of red blood cells to continue for many months after deficient diet or stomach disease deprives the body of fresh supplies.

The formation of red cells occurs in the liver in foetal life, while throughout life their destruction is partly achieved by reticulo-endothelial cells lining its sinusoids.

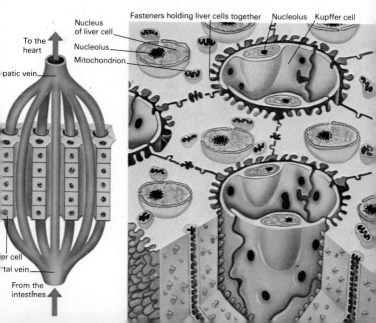

To the heart

patic vein

er cell

tal vein

From the intestines

Nucleus of liver cell
Nucleolus
Mitochondrion

Fasteners holding liver cells together   Nucleolus   Kupffer cell

The pancreas and duodenum

The peritoneum is the shiny transparent membrane which lines the inside of the abdomen and extends over the intestines, providing lubricated surfaces which can glide against each other with next to no friction. Inflammation of the peritoneum produces the painful condition of peritonitis.

The small intestine forms a coiled, twisted mass filling the centre of the abdomen. Unrolled, it is a narrow, mobile tube several feet long in life, but considerably more – 20 feet – after death, when its muscle is flaccid. It is divided into three parts, the first of which, the *duodenum,* lies at the back of the abdomen behind the peritoneum.

A C-shaped tube 10 inches long, the duodenum begins at the pylorus, passes behind the liver, in front of the right kidney and across the aorta. It encircles the head of the pancreas, secretions from which, with bile from the liver, enter it through a duct. The duodenum, like other parts of the

The anatomy of the intestines

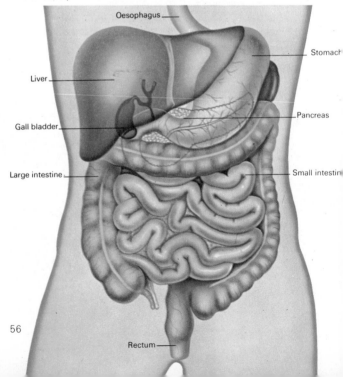

Oesophagus

Stomach

Liver

Pancreas

Gall bladder

Large intestine

Small intestine

Rectum

small intestine, has an outer longitudinal and an inner circular muscle coat. Inside this is the mucous membrane, which has millions of minute finger-like projections called villi, the spaces between which collect the secretions from the epithelial cells. In the duodenum – but not elsewhere – are glands beneath the mucous membrane called *Brunner's glands*.

The pancreas is a fleshy gland 6 inches long, resembling salivary glands in structure and containing, besides the cells secreting pancreatic juice, many tiny *islets of Langerhans* (see page 96-7). 1 to $1\frac{1}{2}$ pints of its fluid are poured daily into the duodenum. It is alkaline and has enzymes which break down starch to maltose (amylase) and fats to glycerol and fatty acids (lipase). It also contains enzymes that are activated in the duodenum and split proteins into polypeptides (trypsin and chymotrypsin). Pancreatic secretion is partly under nervous control but is also stimulated by chemical substances produced in the duodenum when acid enters it from the stomach; this ensures that enough alkali is secreted to reduce gastric acidity sufficiently for duodenal digestion to continue under optimal conditions. Brunner's glands, which produce alkali and mucus, aid this neutralization and also protect the first part of the duodenum from damage by stomach acid.

(*Left*) Magnified section of the wall of the duodenum.
(*Right*) Just as the pleura envelop the lungs, so the peritoneum envelops the continually contracting intestines.

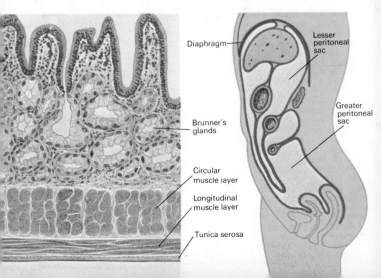

Diaphragm

Lesser peritoneal sac

Greater peritoneal sac

Brunner's glands

Circular muscle layer

Longitudinal muscle layer

Tunica serosa

The small intestine and the large bowel

The second and third parts of the small intestine – the *jejunum* and *ileum* – lead from the duodenum to the large bowel (or large intestine) and are tethered to the posterior abdominal wall by the fan-shaped mesentery. The jejunum and ileum are where the greater part of food digestion and absorption takes place. Food enters the jejunum as a slightly acid liquid and leaves it as an alkaline one; in its passage through, which takes one to two hours, virtually all nutrient material is extracted.

The secretions of the jejunum and ileum contain lipase, maltase, invertase, fructase and amylase (see page 60-1) and activate the pancreatic protein-splitting enzymes. Other enzymes break down polypeptides into amino-acids and split nucleic acids into their components.

Absorption of digestion products begins in the duodenum and is virtually complete by the time the jejunum is reached. This removal of valuable foodstuffs results from effective enzyme action and from the efficient way with which the intestinal lining absorbs them from the cavity of the bowel. The end-products of digestion that are absorbed include glucose (from carbohydrates), aminoacids (from proteins) and a milky fat emulsion.

The large bowel or intestine begins at the caecum, passes upwards and across the abdomen and then down; its name changes from colon to rectum in its last part in the pelvis, and it ends at the anus. The appendix is a blind-ended, worm-like projection from the caecum. It is functionless and often fills with faecal material and becomes inflamed.

The outer longitudinal muscle of the large bowel is arranged in bands. Other obvious differences from the small intestine are a larger diameter and a puckered rather than tube-like outside. The lining secretes mucus but no digestive enzymes, for the function of the large bowel is merely the formation of faeces. These are formed partly from indigestible food material like cellulose, partly from bowel secretions and partly from bacteria. While the upper reaches of the intestines are sterile, the lower parts contain many organisms which manufacture a variety of substances, including vitamin K and the smelly compounds responsible for the odour of faeces. There is little

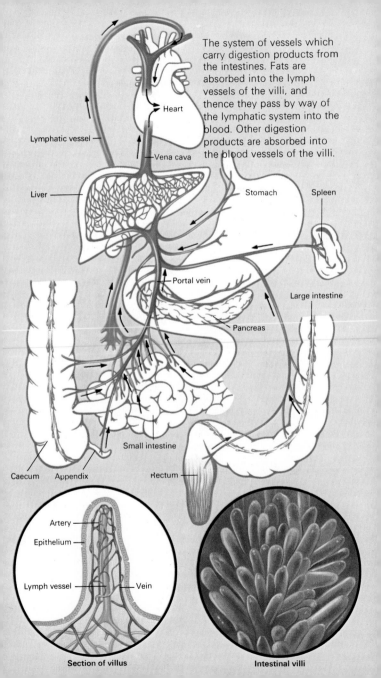

The system of vessels which carry digestion products from the intestines. Fats are absorbed into the lymph vessels of the villi, and thence they pass by way of the lymphatic system into the blood. Other digestion products are absorbed into the blood vessels of the villi.

Heart

Lymphatic vessel

Vena cava

Liver

Stomach

Spleen

Portal vein

Large intestine

Pancreas

Small intestine

Caecum

Appendix

Rectum

Artery

Epithelium

Lymph vessel

Vein

Section of villus

Intestinal villi

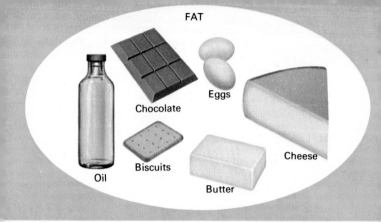

FAT

Oil
Chocolate
Biscuits
Butter
Eggs
Cheese

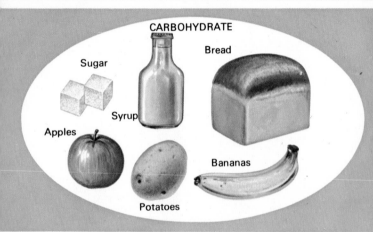

CARBOHYDRATE

Sugar
Syrup
Bread
Apples
Potatoes
Bananas

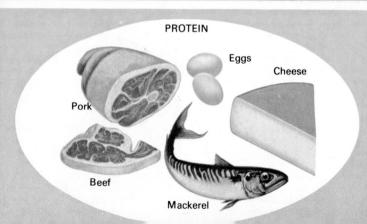

PROTEIN

Pork
Beef
Eggs
Cheese
Mackerel

or no digestible food in the contents of the colon, which gradually become more solid (on account of water extraction) in their 12-hour journey to the rectum. Defaecation occurs when the rectum fills, and is a result of reflex induced peristalsis and relaxation of the anal sphincter.

The sources of energy

To maintain such vital processes as breathing, blood circulation and kidney and brain function, energy is essential. This energy is derived from the three classes of foodstuffs: fat, carbohydrate and protein. *Metabolism* involves the breakdown of these foods with the release of energy and their rearrangement into the complex substances which make up the tissues of the body. Energy release and chemical rearrangement is taking place all the time, and the molecules in almost every tissue or organ are in a state of continual flux.

Carbohydrates make up the bulk of most diets; they are formed of carbon, hydrogen and oxygen with the carbon atoms joined in a ring, and are sugars and starch. The simplest, such as glucose, have only one carbon ring; cane sugar (sucrose) and milk sugar (lactose) have two rings joined together while starch (and also cellulose, which cannot be digested in the human intestine) have many ring units joined together in chains.

Glucose in fact is the carbohydrate which is metabolized to provide energy in the body, and the more complex substances such as sucrose and starch must first be broken down into single units. This is achieved by enzymes in the saliva and small intestine, as has been noted earlier. Ptyalin in saliva breaks down starch in the mouth (this is why a piece of bread chewed for a while begins to taste sweetish) while invertase, lactase and maltase in the small bowel split sucrose, milk sugar and malt sugar each into two single units.

Aminoacid molecules are made of carbon, hydrogen, oxygen and nitrogen arranged in a chain, and *proteins* are composed of many aminoacids linked together chemically. Proteins occur in all tissues and are the basic, indispensible constituents of living cells, of which they form the bulk of the cytoplasm. The digestion of food protein begins in the stomach with pepsin and hydrochloric acid and continues in the duodenum with the action of the pancreatic enzymes. These break down the proteins into *peptide* components and these are further split into their constituent aminoacids, which are absorbed into the blood.

Fats act as a shock absorber, under the skin and around vulnerable organs like the kidneys, and are also a concentrated energy store. Their molecules, containing fatty acids and glycerol, are rich in carbon and hydrogen and so produce many calories when used up in the body. In life the fat under the skin is an oily liquid, but outside the body it becomes a yellowish solid. Fats are digested in the small intestine with the aid of lipase, which disrupts their molecules, and bile salts.

Metabolism

Metabolism is the name given to the biochemical processes occurring when the body builds up living tissue from basic food materials and breaks it down in order to provide energy. It involves the digestion of food in the intestines, the absorbing and storage of the digested materials, their incorporation into the tissues of the body and their release and destruction to water and carbon dioxide with the liberation of energy. This destruction is demanded by the complex chemistry of the body, which requires calories to power its intricate workings. The energy produced by metabolism is not all lost as heat but is instead stored within the cells, in 'energy-rich' compounds which release their calories when needed. These compounds contain special phosphate groupings. The most important is known as ATP (adenosine triphosphate) and when energy is required it releases a phosphate group and a number of calories, becoming ADP (adenosine diphosphate). Later, when it is available, energy is used – and saved up – in converting ADP back to ATP.

Carbohydrate metabolism begins with the absorption of glucose from the hollow of the intestines through their walls and into the bloodstream. Some is carried all over the body and is metabolized by its cells, while some is stored in the liver and muscles as the complex sugar glycogen. This may later be broken down to provide glucose for the body's needs or fuel for muscular activity. When it is needed, glucose enters a 'metabolic pool' in which interplay between the products of the breakdown of carbohydrate, protein and fat takes place.

Aminoacids in the blood stream may come from dietary protein digestion or tissue protein breakdown. Those that are not needed for new tissue formation are metabolized in the 'pool', with urea – which is excreted by the kidneys – as the nitrogenous by-product.

Nucleoproteins are compounds of protein and nucleosides, which are derived from the sugar ribose and nitrogen-containing pyrimidines or purines. The breakdown products of every component save purine enter the 'pool'.

Fat is absorbed into the blood as fine globules and is the major energy store of the body. It accounts for 10 per cent of body weight, and each gram produces nearly twice as many calories as a gram of carbohydrate. Before it can enter the metabolic pool it must be split into fragments containing two carbon atoms.

Metabolic interrelations

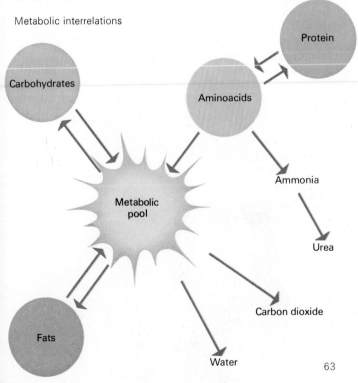

Protein

Carbohydrates

Aminoacids

Ammonia

Metabolic pool

Urea

Carbon dioxide

Fats

Water

63

Diet, energy utilization and requirements

Food must contain adequate water, calories, mineral ingredients and vitamins for health to be maintained.

Water makes up 70 per cent of the body. Some is lost through the kidneys together with waste products, and further amounts are lost in breath, sweat and faeces. This loss must be made good or else dehydration and death will occur.

Energy is required for all living processes; this demands a continuous supply of food. It is only at death that the cells of the body cease to need fuel. During sleep the tissues of the heart, lungs, kidneys and liver continue to work, and the muscles and brain – even though inactive – use up a quantity of energy. The fuel requirements of the body during eight hours of sleep amount to about 500 calories, the *calorie* being the unit of energy used in dietetics. Increasing physical activity obviously demands more energy. Thus writing at a desk uses 40 to 50 calories per hour, light housework needs 100 to 150 calories, and heavy work like road digging uses up to 300 calories per hour. Athletic activities have widely differing calorie requirements, from 150 calories per hour for gentle tennis to 900 calories for a hard game of squash.

Exercise and leisure

Sleep

Sedentary work

Travelling

Clearly optimum calorie intake will vary markedly from person to person, depending on size (in adults the hourly resting requirement is 40 calories per square metre of body surface) and on activity. Growth is something else that needs energy, and despite their size adolescents require at least as great a calorie intake as adults. Correct calorie intake can be worked out knowing that protein and carbohydrate each produce 4 calories and fat 9 calories per gram. A balanced diet should contain at least 60 grams of protein daily and enough carbohydrate and fat to provide the bulk of calorie requirements. In certain conditions it may be necessary to adjust the proportions: in diabetics carbohydrate intake must be regulated, and in kidney failure protein must be cut.

Some minerals, such as cobalt, copper and molybdenum, are needed in minute traces; others, such as sodium and potassium, are important constituents of body fluids. Calcium and phosphate are required for bone formation, iron for haemoglobin and iodine for thyroid hormones.

Calorie requirements vary greatly according to activity.

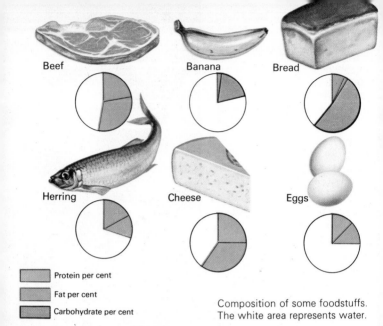

Beef

Banana

Bread

Herring

Cheese

Eggs

Protein per cent

Fat per cent

Carbohydrate per cent

Composition of some foodstuffs.
The white area represents water.

Normal food intake

The idea of a 'normal' diet is valuable in working out on a
large scale the food requirements of a healthy adult population.
In Britain the ideal daily food intake includes 100 grams of
protein, 100 of fat and 400 of carbohydrate. Clearly carbohydrates provide most of the calories, and bread and potatoes
are in this respect the most important parts of the average diet.
Sugar and cereals are also carbohydrate foods. Fats – including
vegetable oils as well as butter, lard and dripping – provide
more energy for the same amount and help to make the
starchy foods more palatable, though they are themselves
unpalatable in quantity.

Food proteins are unimportant as a source of calories but are
vital because some of their component aminoacids – essential
for building new tissues, enzymes and hormones – cannot be
manufactured in the body. They are found in meat, eggs,
cheese and milk as well as certain vegetable products such as
beans, though plant proteins are often deficient in 'essential'
aminoacids. Unfortunately protein foods are expensive, and

poverty unbalances the diet by restricting the protein intake.

Milk is an almost perfect food; although deficient in iron, it contains protein, carbohydrate and fat and is rich in calcium, phosphate and the main vitamins.

Appetite

Exactly what regulates appetite is uncertain, but it is thought that there is a control centre in the part of the brain called the hypothalamus. This centre is sensitive to the amount of sugar in the blood, low sugar levels producing stimulation of appetite. Normally the blood sugar level will vary during the day, producing stimulation or suppression of appetite, but in certain abnormal states it may remain low for long periods and hunger will be experienced. Appetite is of course dependent on many factors, including the appearance, smell and taste of food as well as one's emotional state, illustrating that the hypothalamus is regulated by connections with the cerebral cortex as well as by blood sugar concentration.

Factors influencing appetite

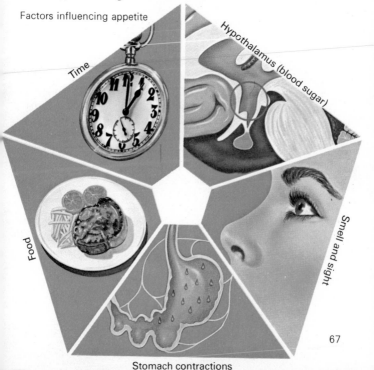

Time

Hypothalamus (blood sugar)

Food

Smell and sight

Stomach contractions

67

Obesity and starvation

Occasionally obesity is due to adrenal overactivity, thyroid deficiency or hypothalamic tumours, but in the vast majority of cases there is no such trouble and by and large the underlying cause of obesity is unknown.

The one certain fact is that *if calorie input is greater than calorie expenditure, weight gain will occur,* but what confuses the problem are the fat people who apparently stick strictly to their diet but are never able to lose weight, and the thin ones who seem to eat excessively but remain underweight. There may be hormonal and hereditary factors in obesity, but these are uncertain. However, if calorie intake is *rigidly* controlled, as occurs in hospital, the obese can achieve weight loss.

This is highly desirable, for obesity leads to heart and kidney disease and shortening of life expectancy. Diabetes is common in the overweight, and osteoarthritis, gallstones and liver trouble are also more likely to occur. Surgical operations are more risky, and the chance of complications in the period after operation is higher.

Perhaps half of the world is undernourished, with a diet

Comparison of diets around the world

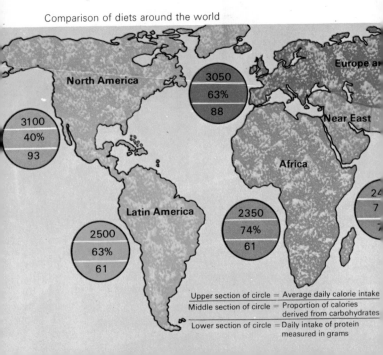

North America

3050
63%
88

Europe an

Near East

3100
40%
93

Africa

2350
74%
61

Latin America

2500
63%
61

24
7

Upper section of circle = Average daily calorie intake
Middle section of circle = Proportion of calories derived from carbohydrates
Lower section of circle = Daily intake of protein measured in grams

lacking both in total calorie content and in protein-rich foods and sometimes deficient in vitamins. The consequences of malnutrition are obvious – stunted growth, a lowered resistance to infectious diseases and a high rate of child mortality. With the population of the world ever increasing and further outstripping its food supplies, this problem is bound to get worse unless new sources of protein are found and there is unprecedented international cooperation.

Malnutrition can have other causes, however. Even with a balanced diet, it may occur if absorption of the digested food is impaired. This happens in a variety of diseases of the intestines, and here malabsorption is associated with steatorrhoea. In the presence of certain chronic infections, and after severe injuries or burns, the body cells may be unable to utilize dietary energy. In widespread cancer, the patient wastes away partly because the multiplying cancer cells eat up much of the available calories and protein.

In starvation – whatever the cause – stores of fat first disappear. Then body protein is used up, the muscles shrink and the patient rapidly loses weight.

(*Right*) Possible effects of obesity

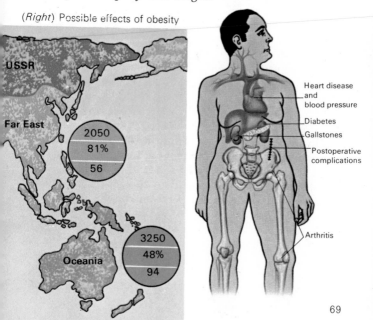

The vitamins

Early in the century it was discovered that animals reared on artificial diets containing adequate water, protein, carbohydrate, fats and minerals, nevertheless failed to develop and grow normally. It was obvious that natural foods contained other substances that were *vital* for health; these were at first thought to be nitrogenous (that is, *amines*). We now know that few vitamins are amines, but the name has stuck.

There are at least a dozen vitamins. They can be divided into two groups, depending on whether they are soluble in fat or in water. The fat-soluble ones are vitamins A, D (anti-rickets factor), E (anti-sterility factor) and K (anti-haemorrhage factor. Vitamins of the B group and C (anti-scurvy) are water soluble.

Vitamins are essential because they are co-factors for enzyme activity; in other words they are necessary aids for the protein catalysts responsible for the processes which go on in the living body. In the absence of vitamins these catalytic systems run down and various metabolic processes may be forced to continue at greatly reduced rates or even to stop. The effects of vitamin deficiencies are thus the result of enzyme deficiencies within the tissues and cells of the body.

Vitamin A

Carotene, which is found in carrots and green vegetables, is a biological precursor of vitamin A and can be converted into vitamin A by the intestinal mucosa. The absorption of both compounds is bound up with factors influencing fat absorption from the gut, particularly the presence of bile salts in the duodenum. Vitamin A itself is present in milk, cheese, butter, eggs and liver and is stored in the body in the liver. It is needed for the regeneration of epithelial tissues, especially the skin, and it is also a precursor of rhodopsin, the light-sensitive substance of the retina. If the dietary intake is inadequate, or if the body is not able to absorb fats from the intestine, deficiency will result. This may cause thickening and dryness of the skin, partial night blindness and in severe cases opacity of the cornea.

Some sources of the principal vitamins

A

Lettuce and spinach

Carrots

Cheese

Cod liver oil

Eggs

Butter

Milk and cream

B

Pork

Mutton

Beef

Liver

Dark bread and cereals

Beans

Peas

Cabbage

C

Oranges

Tomatoes

Lemons

Red and green peppers

Green vegetables

Black currants

Strawberries

Potatoes

D

With sunlight, the body makes vitamin D from a substance found in the skin.

Cod liver oil

Cabbage

Eggs

Butter

Milk and cream

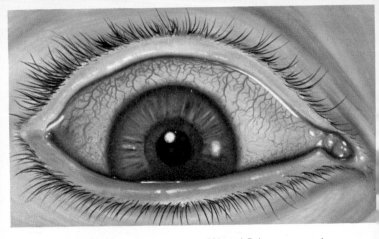

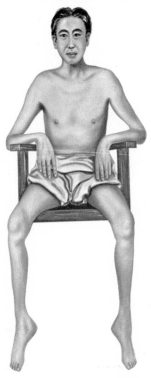

(Above) Enlargement and spread of blood vessels of the conjunctiva in riboflavine deficiency. (Left) 'Dry' beri-beri resulting from thiamine deficiency. Wrist drop, foot drop and wasting of lower extremities are present.

The B vitamins

The B vitamin group includes thiamine (B_1), nicotinamide, riboflavine, pyridoxin (B_6), folic acid and cyanocobalamin (B_{12}). The first three are vital factors in carbohydrate metabolism and oxidative reactions; the fourth is involved in aminoacid metabolism, and the last two in blood formation.

Thiamine is found in beans, peas, cereals and yeast, in bran, wheat germ and rice husks and to a small extent in milk and meat. It can survive a short spell of boiling.

Thiamine deficiency causes disease of the heart, nerves and brain. The function of the heart deteriorates in its absence, and eventually heart failure occurs. One sign of this is marked ankle swelling, due to collection of fluid in the tissue under the skin (oedema). This gives the condition the name 'wet' beri-beri. The heart failure clears up when thiamine is given. In 'dry' beri-beri the nerves are affected, with muscle weakness and impaired feeling in the hands and feet; recovery is slow and often incomplete after treatment, indicating that there has been permanent damage. Brain involvement from thiamine deficiency usually occurs in poorly nourished alcoholics, and produces unsteadiness, loss of memory, drowsiness and coma. It may be fatal if it is not treated promptly.

Nicotinamide, which is not destroyed by cooking, is present in liver, yeast, meat and green vegetables. Deficiency causes undue sensitivity to sunlight, with severe skin damage, and also affects the cells lining the intestine and the cells of the brain, producing diarrhoea and mental changes. This condition is called pellagra.

Riboflavine is obtained from meat, milk and whole-meal flour. It is relatively heat-stable but is destroyed by light. Riboflavine deficiency has no life-threatening effects. It causes changes in some epithelia, and this is most marked in the tongue, which becomes smooth, sore and purplish.

(*Left*) Pellagra dermatitis resulting from nicotinamide deficiency.
(*Right*) Soreness and swelling of tongue and lips in riboflavine deficiency.

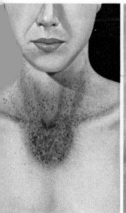
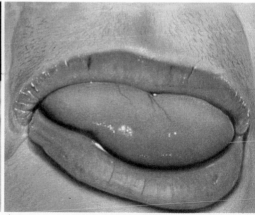

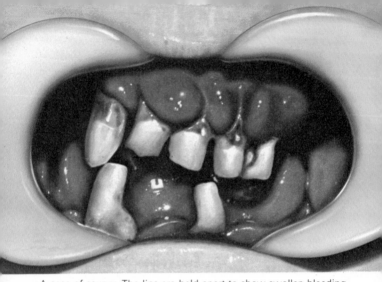

A case of scurvy. The lips are held apart to show swollen bleeding gums and loosening of the teeth.

Vitamin C

For over 200 years it has been known that scurvy is due to a deficient diet and can be cured by fresh fruit and vegetables, but the anti-scurvy factor, vitamin C (ascorbic acid), was only discovered in 1932. Citrus fruits, brussels sprouts, black currants, tomatoes and cabbage are good sources. Human milk (but not cow's milk) is another source, but there is no vitamin C in meat. It is destroyed by cooking, and so it is usually absent in preserved or canned foods.

Vitamin C is essential for the formation and repair of the connective tissue and the ground substance that surrounds cells, as well as for the repair of capillary blood vessels. Deficiency is a result of inadequate intake, due to poverty, an unbalanced diet, or both. It often occurs in old people who live alone and sometimes in babies who are fed on heated cow's milk without fruit juice supplements. Because capillary walls are affected, patients with scurvy bruise readily and their gums bleed and swell. Skin wounds heal slowly. In infants, bleeding may occur into the intestines and around bones, and the bones themselves grow slowly because of lack of cement substance.

Vitamins D, E and K

Vitamin D is found in eggs, butter, milk, fish and fish liver oil, but not in vegetables. It is a lipid substance related to cholesterol, and one variety of the vitamin is formed in the skin by the action of sunlight on 7-dehydrocholesterol.

Its action is to increase the absorption of calcium and phosphate from the intestines and it is essential for the laying down of calcium bone. Its deficiency causes rickets; this may occur if the diet is inadequate, and in city dwellers who are cut off from bright ultraviolet light. The effects in children are a result of an insufficient calcium supply to growing bones. Growth is stunted and weight-bearing parts of the skeleton like the legs and spine become curved. Teeth develop badly and there may be muscular weakness. Rickets is today prevented by free school milk and adequate basic nutrition. Ultraviolet light and cold liver oil provide further supplies of vitamin D.

Vitamin E, which prevents abortion in rats, has no such obvious effect on humans. Vitamin K is needed for prothrombin formation (page 117). Small quantities are present in many foods, and deficiency results not from inadequate diet but from disease of the liver, biliary system and intestines.

(*Left*) X-ray of a case of rickets. (*Right*) Child with rickets

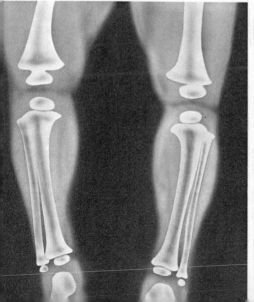
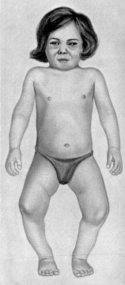

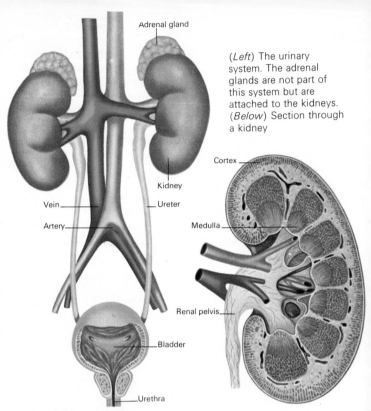

(*Left*) The urinary system. The adrenal glands are not part of this system but are attached to the kidneys. (*Below*) Section through a kidney

Adrenal gland

Kidney

Vein

Artery

Ureter

Cortex

Medulla

Renal pelvis

Bladder

Urethra

THE URINARY SYSTEM

The anatomy of the urinary tract

The kidneys are on the back wall of the abdomen, just below the diaphragm and on either side of the vertebral column. Each one weighs about 5 ounces. It is supplied with blood by a single artery which comes from the aorta and divides into three branches. These are accompanied by veins which join to form the renal vein. This drains blood into the inferior vena cava.

Urine is formed in the substance of the kidneys and collects in the thin walled, funnel-like renal pelvis. This leads into the ureter, the muscular wall of which contracts in slow waves, thereby helping the urine along to the bladder. Here urine is stored – a pint or so can be held with ease. In micturition the tight sphincter at the base of the bladder relaxes and the smooth muscle wall of the bladder contracts. Although this action is really involuntary, it is well controlled by the cerebral cortex.

In cross-section, the kidney is composed of an outer cortex and an inner medulla. The cortex contains many thousands of tiny glomeruli; together with the renal tubules, which partly lie in the medulla, they make up the *nephrons* or functional units of the kidney.

Glomeruli are small knots of capillaries, derived from the finest branches of the renal artery. These knots are encased by Bowman's capsule, which is the bulbous ending of the cell-lined tubule. After a short twisted course, the tubule passes down into the medulla towards the renal pelvis and then doubles up on itself, travels up to the cortex where it twists again before its final downward passage to join the collecting system terminating in the renal pelvis.

The excretory unit of the human kidney. Thousands of these units make up each kidney.

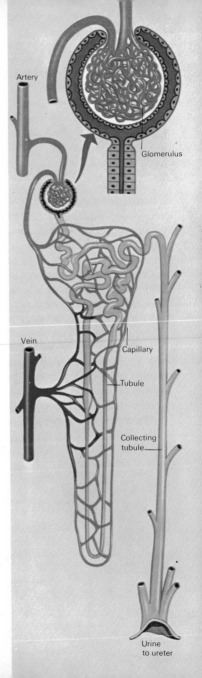

Artery

Glomerulus

Vein

Capillary

Tubule

Collecting tubule

Urine to ureter

The functions of the kidneys

The major function of the kidneys is the removal of waste products, with the formation of urine. This is achieved while controlling the amount of salt and water in the body and maintaining the slight alkalinity of body fluids.

Cell metabolism produces various by-products which might be lethal if they were allowed to accumulate in the tissues. They are removed by the nephrons, which act as filters. From the blood in the glomeruli, fluid oozes into Bowman's capsule. Blood cells and large protein molecules remain in the capillaries because the pores in the 'filter' are too small to let them out. The filtered fluid then passes through the first downward part of the tubule, where 80 per cent of the water and sodium in it is reabsorbed by the lining cells, and travels up into the cortex again. Here, under the influence of the hormone aldosterone, the remaining sodium is removed from it, and

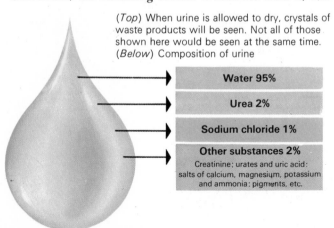

(*Top*) When urine is allowed to dry, crystals of waste products will be seen. Not all of those shown here would be seen at the same time.
(*Below*) Composition of urine

Water 95%

Urea 2%

Sodium chloride 1%

Other substances 2%
Creatinine; urates and uric acid; salts of calcium, magnesium, potassium and ammonia; pigments, etc.

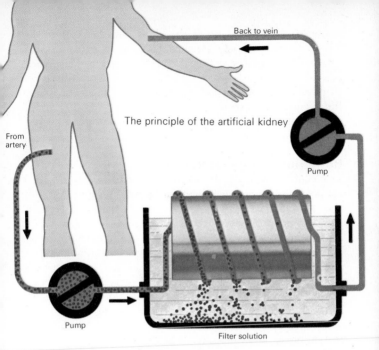

The principle of the artificial kidney

Back to vein

From artery

Pump

Pump

Filter solution

hydrogen ions are excreted in exchange by the lining cells. The acidic liquid thus produced travels into the collecting ducts, the cells of which, stimulated by antidiuretic hormone, take up almost all the water that is left. Urine, now acidic and highly concentrated, drains into the renal pelvis and passes via the ureter to the bladder.

The kidneys – which receive one fifth of the cardiac output – filter off through their glomeruli 120 ml of fluid every minute. They are very efficient in retaining water, for only 1 ml of this becomes urine. In fact just the minimum quantity needed to dissolve waste products is allowed to escape from the body.

Kidney damage and the artificial kidney

The kidneys are very susceptible to lack of oxygen and a severe fall in blood pressure, which stop them functioning and damage their cells. Renal failure may also occur more gradually in a variety of diseases which affect the glomeruli and tubules. Whatever the cause, the effect is the same – toxic and acid waste products build up in the blood, and the patient dies unless treated. Kidney transplantation is possible in some cases; in others the artificial kidney is used. This works basically by osmosis – impurities in the blood are filtered off by passing it through a coiled tube immersed in a bath of clean fluid. The tube is a semipermeable membrane, allowing soluble substances of small molecular size to pass out while keeping in larger molecules.

Body fluids and electrolytes

The main constituent of the human body is water, which accounts for nearly two-thirds of its total weight. An adult man of average weight – 70 kilograms – therefore contains about 45 litres of water (a woman of the same weight would contain rather less as she has a greater amount of fat). Of this volume, about 30 litres is within the cells, maintaining their size and shape and acting as a vehicle for movement of salts, glucose, oxygen and the products of metabolism. About 3 litres of water circulates in the heart and blood vessels. Since a little over half of blood is plasma (the rest being red and white cells), the total blood volume is roughly twice this. The remaining 12 litres of water is extracellular, and lies in the space between cells. The ability of the kidneys to retain the water of the body is obviously very important, for if all the

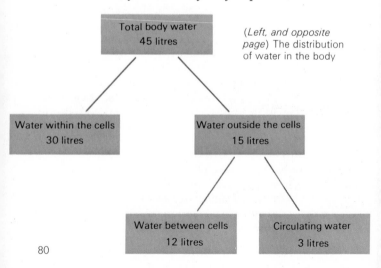

(*Left, and opposite page*) The distribution of water in the body

Total body water
45 litres

Water within the cells
30 litres

Water outside the cells
15 litres

Water between cells
12 litres

Circulating water
3 litres

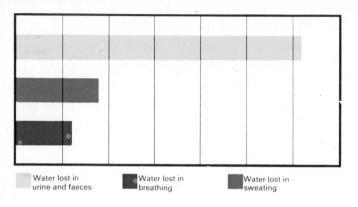

Water lost in urine and faeces

Water lost in breathing

Water lost in sweating

fluid filtered by the glomeruli were to be excreted, the plasma volume would be lost within half an hour.

The salts dissolved in the body fluids – the electrolytes – are in the form of ions. Sodium, chloride, bicarbonate, potassium, calcium and phospate are the ions present in greatest quantity. An important fact is that while outside the cells sodium is found in high concentration and potassium in low concentration, the reverse holds true inside the cells. Potentially there is free movement of different ions in and out of cells, at a rate which depends on their relative concentration in the two sites. The distribution of sodium and potassium in the body is thus the result of active physico-chemical processes; these two ions are segregated by the cell membranes, and this segregation is vital for cell function.

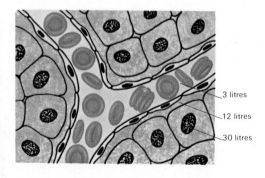

3 litres

12 litres

30 litres

HORMONES AND REPRODUCTION

Some glands, such as those of the skin and the gastrointestinal tract, produce secretions which drain on to the surface or into a body cavity and have an effect only near where they are released. These are called *exocrine* (*exo* = outside). In contrast, the *endocrine* glands (*endo* = inside) manufacture substances which pass directly into the bloodstream, circulate all over the body and act at places far from where they originate. Such substances are *hormones*; at present nearly thirty are recognized, the products of eight different glands. They are mostly fairly simple organic chemical compounds and some have been made in the laboratory from artificial ingredients. They are effective in very small amounts and are the regulators of body processes, controlling growth and development, sexual activity, pregnancy and birth, and metabolism. Only in this century had hormone action begun to be understood, though before this the effects of damage to one or other of the endocrine glands were known – removal of the testes resulting in the excessive fat and sexual impairment of a eunuch; damage to the pancreas producing diabetes; shrinkage of the adrenals causing Addison's disease; and so on.

Regulation of body processes depends on regulation of hormone production. A few of the endocrine glands control themselves, but for the most part hormone secretion is under the control of the pituitary, a bean-shaped gland lying in a little cavity in the centre of the base of the skull. The front or anterior part secretes a number of substances which go out in the blood to other glands and stimulate hormone formation; these hormones then circulate round to the pituitary and, if too much has been formed, suppress the stimulator – there is thus a sort of feedback system. The adrenal cortex, thyroid and sex organs are dependent on the pituitary. If this is damaged, these glands shrink and are unable to function.

The pituitary gland

Under the microscope the *anterior pituitary* is seen to be made up of nests of cells separated by blood vessels and connective tissue. The cells are of three sorts: acidophil (taking up, and thus stained by acid dyes), basophil (stained by basic dyes)

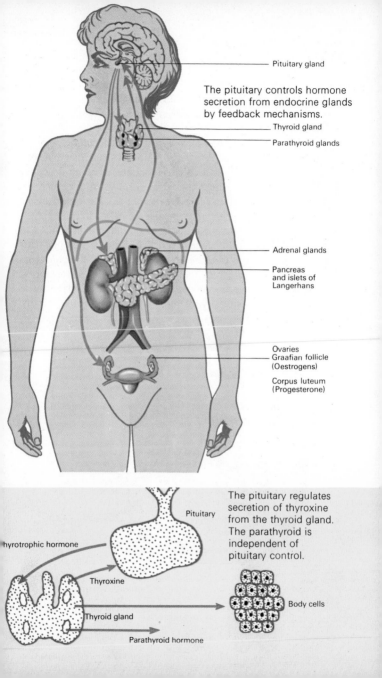

Pituitary gland

The pituitary controls hormone
secretion from endocrine glands
by feedback mechanisms.

Thyroid gland

Parathyroid glands

Adrenal glands

Pancreas
and islets of
Langerhans

Ovaries
Graafian follicle
(Oestrogens)

Corpus luteum
(Progesterone)

The pituitary regulates
secretion of thyroxine
from the thyroid gland.
The parathyroid is
independent of
pituitary control.

Pituitary

hyrotrophic hormone

Thyroxine

Body cells

Thyroid gland

Parathyroid hormone

and chromophobe (without special affinity for dyes). Usually these stain red, blue and grey respectively. The acidophil cells produce growth hormone and also a hormone which stimulates milk production in pregnancy. The basophils secrete hormones which stimulate the thyroid, adrenal cortex, ovaries and testes. Some chromophobes secrete hormones while some are probably inactive forms of the other cell types.

With pituitary tumours, excess growth hormone may be manufactured. If this occurs before puberty, there is gigantism because the bones are still able to elongate; after adolescence the growing ends of the bones (the epiphyses) are inactive, and the hormone causes the bones to get thicker and the soft tissues of the body to enlarge. A disease called acromegaly is the result. Insufficient growth hormone may be produced in pituitary disease, and some dwarfs are so because of this occurring before adolescence. In adults pituitary damage from any cause is likely to result in depletion of all its hormones, leading to loss of weight and depression of thyroid, adrenal and sex functions. Occasionally pituitary damage is due to pressure

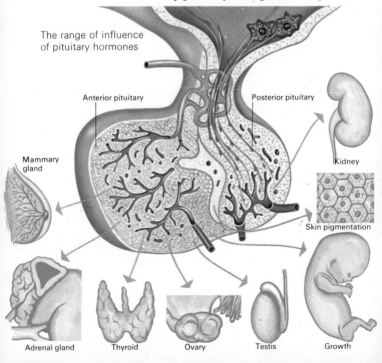

The range of influence of pituitary hormones

Anterior pituitary

Posterior pituitary

Mammary gland

Kidney

Skin pigmentation

Adrenal gland   Thyroid   Ovary   Testis   Growth

from a tumour of the basophil cells, with increased production of ACTH (the adrenal cortex stimulating hormone) and signs of adrenal overactivity.

The *posterior pituitary* is really an extension of the brain, unlike the anterior part which is derived from the palate of the embryo. It contains nerve fibres, connective tissue and blood vessels. Its hormones are produced in the adjoining hypothalamas and are passed along nerve fibres in the stalk of the gland to be stored in the pituitary. There are two, antidiuretic hormone (ADH) and oxytocin. As noted on pages 78-9, ADH stimulates the kidney tubules to reabsorb water from the urine, and in damage of the posterior pituitary, *diabetes insipidus* results (insipid – as distinct from *mellitus*, sweet). In this condition 15 times the normal urine volume, or more than 30 pints a day, may be produced; the patient has to drink water by the bucketful to keep up with his kidneys.

Oxytocin causes contraction of the muscles of the viscera and thus helps the uterus to function in childbirth. It also causes ejection of milk from the breast in suckling.

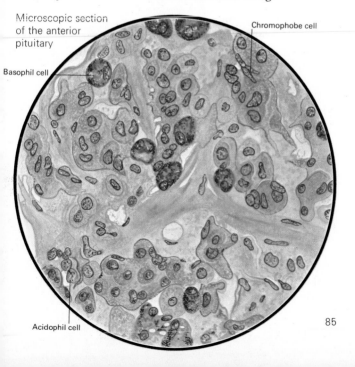

Microscopic section of the anterior pituitary

Chromophobe cell

Basophil cell

Acidophil cell

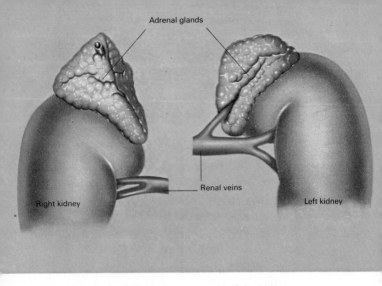

The adrenal glands lie at the upper end of the kidneys

The adrenal glands

Like the pituitary, the adrenal gland is a double organ, with two distinct and separately functioning parts – the outer *cortex* and inner *medulla*. There are two adrenals, yellow-brown triangular glands lying close to the upper end of the kidneys alongside the spine, and together they weigh about 1/4 ounce. Like all endocrine glands, they have a rich blood supply. A microscope section shows that the cortex is made up of lightly staining cells containing much fat, arranged in three zones. Outside is a zone where the cells are in rounded groups, and beneath this there is a layer of cell columns running at right angles to the surface and an innermost irregular layer.

The *adrenal medulla* contains columns of large cells that manufacture *adrenaline* and *noradrenaline,* the two hormones that produce the effects of the *sympathetic nervous system.* There are cells of this system in the hypothalamus and spinal cord and in a chain running alongside the vertebrae, and its nerves go to the heart, blood vessels, intestines, bladder and wherever else there is smooth muscle, releasing noradrenaline at their endings. This hormone causes quickening of the pulse, increased blood pressure, constriction of blood vessels, decreased activity in bladder and bowel muscle, dilation of the pupil and a rise in blood sugar – all in preparation for 'flight or fight'. Noradrenaline and adrenaline from the adrenals enter the blood to increase these actions. The effects of the two hormones differ in a number of important ways, and they are probably produced by two different sorts of cells. Thus of the two, adrenaline had much stronger inhibitory actions (for example on the bladder and intestines), while noradrenaline is more powerful in raising the blood pressure and constricting arterioles.

When the sympathetic nervous system is active its effects are outwardly obvious, for the skin is pale (due to vasoconstriction) and there is gooseflesh (due to stimulation of the smooth muscle attached to the base of the hairs). There is also sweating, which however is produced not by adrenaline but by *acetylcholine*. This substance is involved in the transmission of nerve messages to voluntary muscle, but it also has effects on smooth muscle which are more or less opposite to those

Adrenal gland section (*left*) and 'flight or fight' reaction (*right*)

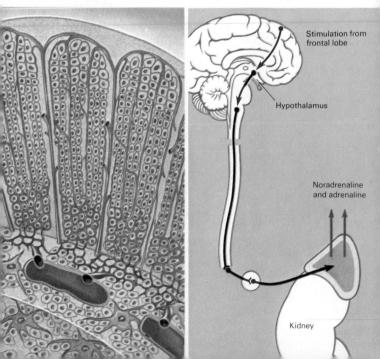

Stimulation from
frontal lobe

Hypothalamus

Noradrenaline
and adrenaline

Kidney

produced by the adrenal hormones. It is in fact the agent of the *parasympathetic nervous system* and is released from the parasympathetic nerve endings in smooth muscle all over the body.

The *adrenal cortex* forms a number of hormones which are essential to life. There are the *corticosteroids,* and the two most important are *aldosterone* and *hydrocortisone.* Respectively they regulate the quantity of salt and water in the body, and control carbohydrate, protein and fat metabolism and help fight stressful conditions. Several other corticosteroids are produced in the adrenal, including cortisone, which is nowadays synthesized and used in treatment.

The actions of aldosterone and hydrocortisone are clearly seen in states of adrenal under or overactivity. Addison's disease is due to diminished adrenal function; patients with this condition suffer from weakness and tiredness, wasting (because of the lowering of blood sugar and lack of formation

Cushing's disease

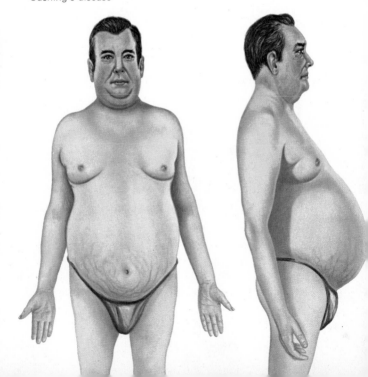

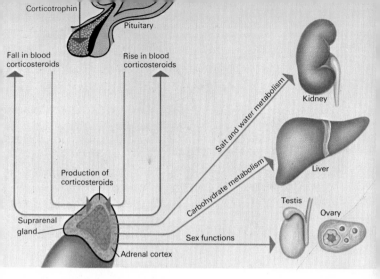

Control and influence of the adrenal cortex

of new tissues), low blood pressure (because of lack of salt and water and decreased blood volume), loss of body hair, and sensitivity to stresses of various sorts. In addition they become pigmented, because pituitary ACTH secretion increases to try to make up for the low blood level of corticosteroids, and ACTH also stimulates the melanin-forming cells of the skin. Increased cortical activity – often due to an adrenal tumour but occasionally seen with a pituitary tumour – produces *Cushing's disease*. Here the patient has thin limbs (because muscle protein is broken down to sugar), a fat trunk and rounded face (sugar is converted into fat), excessive hairiness (overproduction of male hormones), diabetes (high blood sugar, with insulin production inadequate to satisfy demands) and high blood pressure (excess aldosterone causing salt and water retention and increased blood volume).

Apart from cortisone, other corticosteroids have been synthesized, including some artificial ones which are more powerful than aldosterone and hydrocortisone. These include prednisone, which is used in inflammatory conditions, and fluorocortisone, which has a potent salt-retaining effect. Corticosteroids are also used to suppress immune reactions.

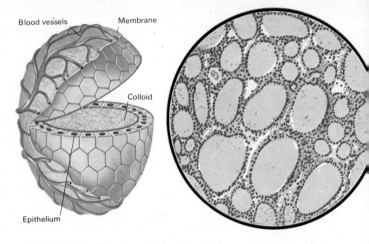

Blood vessels

Membrane

Colloid

Epithelium

The position of the thyroid gland in relation to the larynx and trachea is shown opposite. The structure of a follicle is shown above (*left*) and a microscopic section through the thyroid gland (*right*).

The thyroid gland

The thyroid is a fleshy pink-brown gland lying in the neck on either side of the larynx and weighing about one ounce. Two lobes are connected across the front by a narrow isthmus. Thryoid tissue has a rich blood supply, and several arteries enter its upper and lower parts. Four small *parathyroid glands,* each about the size of a matchhead and yellowish brown in colour, are loosely attached behind the upper and lower ends of the thyroid, but sometimes the lower two are much further down and in the chest.

The surface of the thyroid is glistening and granular, for the gland is made up of tiny spherical follicles containing a jelly-like material. The epithelium of the follicles manufactures thyroid hormones which are then attached to protein and stored as a gel. When the gland is hyperactive, the hormones are released and the follicles collapse. An essential component of these hormones is iodine, which must be present in the diet in adequate amounts for the gland to function normally. Where there is insufficient iodine – especially in mountainous parts of the world – the gland enlarges in an attempt to compensate and a goitre results. The signs of *hypothyroidism*

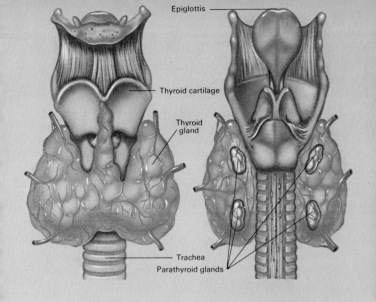

Epiglottis

Thyroid cartilage

Thyroid gland

Trachea

Parathyroid glands

will occur if there is a marked dietary insufficiency of iodine.

The chief thyroid hormone is called thyroxine, and each molecule contains four atoms of iodine. Tri-iodothyronine has one atom less: it is produced in smaller quantity but is more powerful. The two hormones together stimulate metabolism and heat production, and if too much of them is produced this stimulation becomes excessive and the condition of thyrotoxicosis is obtained.

In *thyrotoxicosis,* which occurs most often in young women, the metabolic rate is very high. Carbohydrates, fats and proteins are needed in excess of normal and so, despite a good appetite and large food intake, the patient tends to lose weight. Subcutaneous tissue disappears and muscle bulk decreases. People with thyrotoxicosis are hot and sweaty, with an overactive circulation and fast pulse. There is usually a slight but noticeable swelling of the gland and facial appearance is altered by the protrusion of the eyes. In young people the condition is treated by removing most of the over-functioning thyroid, but in older patients a dose of radioactive iodine is given – this is concentrated in the gland and the radioactivity kills the hormone-producing cells.

A smaller quantity of radioactive iodine is used in the diagnosis of thyroid diseases, for the amount concentrated in the gland in the neck can be measured. An overactive gland absorbs more than usual, and vice versa.

Underactivity of the thyroid in the adult causes *myxoedema* and in infancy *cretinism*. Thyroxine is not only a metabolic stimulator but also an essential substance for body growth and the development of the brain. Cretins, who are thyroid deficient from birth, are stunted and severely retarded mentally. They have a coarse skin and features and a large tongue. Cretinism may result from iodine deficiency or from the absence of enzymes normally present in the thyroid. Treatment with thyroxine improves the general state but influences the mental condition only if it is started very early in life. In adults *myxoedema* is a condition in which patients are cold, slow physically and mentally, with a puffy face and swollen limbs and body, due to the laying bown beneath the skin of a mucinous fluid. They feel excessively cold in winter, and as they lack the substances which stimulate metabolism they are

Myxoedema (*left*) and thyrotoxicosis (*right*)

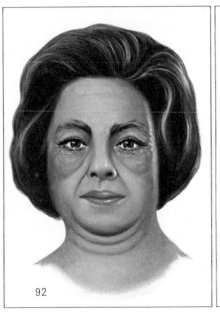
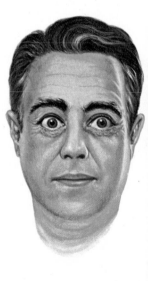

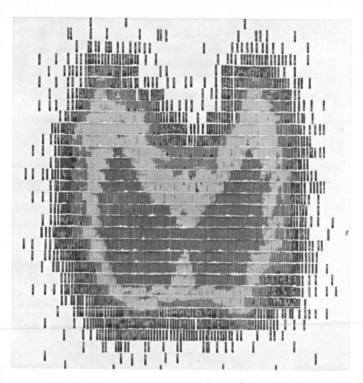

One type of radioactive iodine scanner plots the iodine content of the thyroid in colours according to local concentration, from red (high) to blue (low).

prone to *hypothermia,* or marked lowering of body temperature, with severe consequences. Patients with myxoedema also tend to get anaemia (thyroxine being essential for the normal formation of red blood cells) and heart disease, so diagnosis and treatment with thyroid hormones is vital.

An entirely different sort of hormone is made in the thyroid gland. This is *calcitonin,* which is secreted not by the follicular cells but by cells lying between the follicles and coming from a different source in the embryo. Its action is to lower the level of calcium in the blood, and it has thus the opposite effect to the parathyroid hormone.

The parathyroid glands

The parathyroids under the microscope can be seen to consist of cords of cells separated by wide blood spaces. There are two types of cells, most having a clear cytoplasm and a faint nucleus while a minority are red-staining with well-defined nuclei. The hormone secreted by these glands is called *parathormone*, and it stimulates the release of calcium from bone – by stimulating osteoclasts – while increasing the excretion of phosphate from the kidneys.

In the body, calcium and phosphate ions go hand in hand, an increase in one causing a reciprocal decrease in the other. Bone consists basically of a connective tissue scaffolding supporting a hard mass of mineral crystals, which are mainly calcium phosphate. When the parathyroid glands are overactive these crystals are removed from bone. Calcium and phosphate are poured into the bloodstream, but phosphate is selectively excreted in the urine, so that its level stays low. Some calcium is also lost from the kidneys, but the blood level remains high. As a result of the loss of calcium phosphate, the bones become soft and brittle. They may fracture with the smallest strain or they may twist and bend, as occurs in the condition called *osteitis fibrosa cystica*. Bone disease is not the

Microscopic section through a parathyroid gland

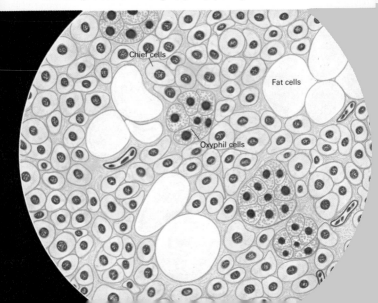

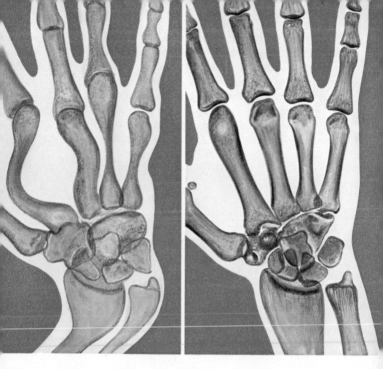

X-rays to show condition of osteitis fibrosa cystica (*left*) and appearance of normal limb (*right*)

only consequence of parathyroid overactivity, for the freed calcium in the blood is deposited in a variety of tissues, impairing their function. This is especially important in the kidneys, which may be severely involved; the pancreas is also sometimes affected, with painful inflammation.

Underproduction of parathormone has the opposite effect – a lowered blood calcium. This happens after the accidental removal of some of the parathyroid glands (a hazard of thyroid surgery), and its most important feature is tetany, or spontaneous prolonged twitching of muscles. Lowering the blood calcium level makes all the muscles abnormally excitable, so that the hands, feet and face may twitch distressingly. An injection of calcium cures this at once; administration of parathyroid extract has the same effect.

Diabetes and the islets of Langerhans

Scattered throughout the pancreas, between the glands that pour juice into the pancreatic duct, are many pinhead-sized groups of special cells. These are the islets of Langerhans and they secrete the hormone *insulin*, a deficiency of which is responsible for diabetes mellitus. With suitable stains, two types of islet cells can be demonstrated, each with characteristic granules. It is the beta cells which manufacture insulin and release it into the bloodstream.

A chemist's chart showing colour coding for cartons of insulin, indicating the types and strengths available

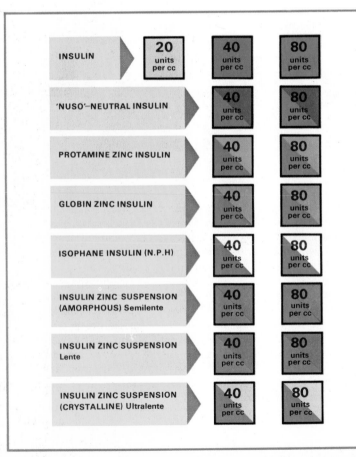

Insulin is a protein and its chief effect is to produce a lowering of the amount of glucose ('sugar') in the blood. It does this in several ways – by increasing the glycogen stores in the liver and the muscles, by hastening the entry of glucose into the cells and stimulating its breakdown into water and carbon dioxide, and by promoting fat deposition. If insulin is deficient, the whole metabolism of the body suffers, though the immediate and obvious effect is an increase of blood glucose, which appears in the urine (hence diabetes *mellitus* – sweet). Insulin lack also causes an excessive breakdown of fats, so that strongly smelling ketones are produced. By osmotic action, the glucose in the urine takes with it a lot of water and also salts, and so someone with severe diabetes is thin and dehydrated, with ketones on his breath and sugar in his urine, and will go into coma and die unless treated. Before insulin was isolated about fifty years ago, this disease was always fatal. Now it is fairly easily controlled, the patient injecting himself with the required dose of insulin. Different sorts of insulin with varying lengths of action are available, as well as tablets containing substances which lower the blood sugar

Microscopic section of the pancreas

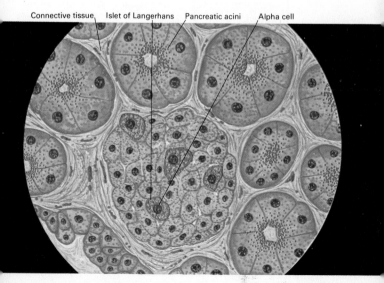

Connective tissue   Islet of Langerhans   Pancreatic acini   Alpha cell

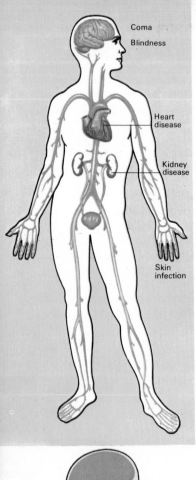

Coma
Blindness
Heart disease
Kidney disease
Skin infection

level – these however cannot replace insulin if the diabetes is at all severe. Some patients with mild diabetes can be controlled with tablets or with dietary carbohydrate restriction, but moderate or severe diabetes always needs insulin.

Diabetes entails not only a disturbance of metabolism but also structural changes in the kidneys, eyes, blood vessels and nerves. These can cause kidney failure, blindness and gangrene of the feet, but can be prevented by careful treatment and regular medical checks.

The alpha cells of the islets secrete a hormone called *glucagon,* which raises the blood glucose level by stimulating the breakdown of liver glycogen. The function of glucagon in the body is uncertain, but it is of minor importance compared with insulin.

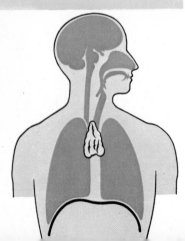

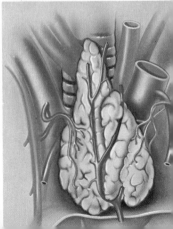

The thymus

The thymus has not yet been shown to secrete a hormone but it nevertheless exerts a powerful influence on the whole body, especially in the early part of life. It is a fatty bilobed gland at the root of the neck and in the front part of the chest. Its outer part contains millions of dividing lymphocytes while the inner zone contains primitive cells and structures called *Hassall's corpuscles*. At birth it is fairly large. It continues to grow up to the time of adolescence, after which it shrinks. By middle age all that remains is a few strands of tissue.

The thymus is of the utmost importance in the body's defences against infection, for in the first few weeks of life the lymphocytes produced in it migrate into the bloodstream and colonize lymph nodes all over the body. Lymphocytes manufacture antibodies and are vital for immunity. If the thymus of newborn animals is damaged, there is a marked stunting of growth, development is retarded and there is a great susceptibility to infections – due to a reduction of the number of lymphocytes and impairment of antibody formation. The reduced resistance to infection which is brought about by corticosteroid treatment is paralleled by a shrinkage of the thymus and lymphoid tissue generally. This will lead ultimately to a lowered capacity for antibody formation.

(*Opposite page*) Those parts of the body affected by an acute deficiency of insulin.
The position of the thymus in the neck (*far left*), its blood supply (*left*) and microscopic structure (*below*)

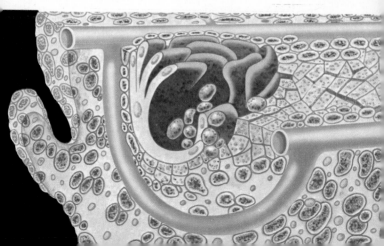

Determination of sex

Whether a being is male or female is already determined at the time of conception. All human cells have 23 pairs of chromosomes, one of these pairs being the sex chromosome. In women the sex chromosomes are an identical pair (XX), but in men one of the two is smaller (XY). When reduction division (page 6) occurs in the germ cells of the ovaries and testes, the pairs are split. Thus all ova will have 22+X chromosomes, as will half the spermatozoa, but the other half will have 22+Y, and when one of these unites with a 22+X ovum a male embryo results.

Occasionally reduction division goes wrong, and instead of having XX or XY sex chromosomes the fertilized ovum may have an extra X or a deficient Y. In such cases, characteristics of both sexes may be present. Intermediate sexuality can also occur as a consequence of abnormal hormone production in early life (thus overactive adrenal glands may produce external male features in baby girls) while in adults, male or female, secondary sex characteristics may be acquired inappropriately as a result of hormone therapy.

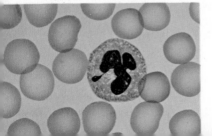

Human sex determination
Nucleus with sex chromatin

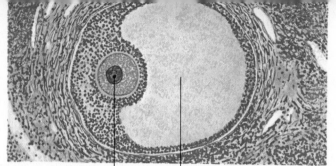

Germinal vesicle   Follicular fluid

Microscopic section of an ovarian follicle

The ovaries, ovulation and menstruation

The female equivalents of the testes lie inside the pelvis. The size of a bean, they are supported by ligaments and are touched by the fronded ends of the Fallopian tubes, the ducts for the passage of ova to the uterus. The ovaries are the source of female germ cells (ova) and the female hormones oestrogen and progesterone. Ova are produced from follicles. These are minute spherical structures, precursors of which are present from infancy, but which begin to mature only at puberty. While up to 400,000 precursors are present in childhood, only one a month matures between the onset of menstruation and the time of menopause – a total of perhaps 400.

The maturing ovarian follicle secretes *oestrogen*, which enters the bloodstream. Release of the ovum from the follicle occurs in most women in the second week after a menstrual period. After release, the wall of the mature follicle thickens and its cells multiply, forming the *corpus luteum*, which manufactures *progesterone* as well as oestrogen.

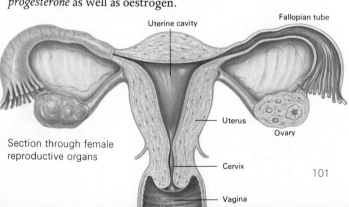

Uterine cavity

Fallopian tube

Uterus

Ovary

Cervix

Section through female
reproductive organs

Vagina

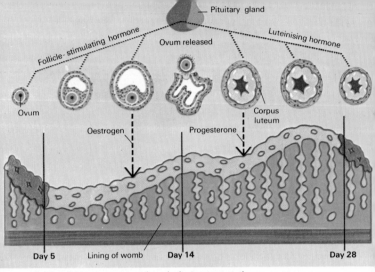

Cycle of hormone secretion during menstruation

Menstruation, the shedding of the glandular lining of the uterus, is the result of cyclical change of hormone secretion. Under the influence of oestrogen alone (that is, in the first part of the month, before ovulation) the lining thickens, its capillaries multiplying and glands enlarging. Then, when progesterone is secreted, greater thickening occurs. Just before menstruation the corpus luteum degenerates; the supply of hormones is cut off and the uterine lining starts to slough. This continues for a few days by when another follicle is maturing and a new cycle has begun.

Ovarian hormone production is under the control of the pituitary, which triggers sexual development at puberty. One pituitary hormone (follicle-stimulating hormone) is active in the first half of the cycle; increasing blood oestrogen suppresses it and elicits secretion of luteinising hormone, which activates the corpus luteum.

Pregnancy

After ovulation the ovum passes along the Fallopian tube and into the uterus. Unless fertilized, it dies within two days. However, if intercourse has taken place during this time, or within two days before ovulation, there is a chance that it will encounter a spermatozoon. When this happens, the fertilized ovum divides and becomes implanted in the thick glandular lining of the uterus. A placenta soon forms in this lining and produces a hormone called *chorionic gonadotrophin*. This prevents the disappearance of the corpus luteum and indeed stimulates it to secrete more oestrogen and progesterone. The build-up of hormones is responsible for many of the physical changes of pregnancy – enlargement of breasts and uterus, laying down of fat, and fluid accumulation.

The developing fertilized ovum is an embryo in its first eight weeks and a foetus thereafter. It is nourished by the placenta, in which exchange of foodstuffs and oxygen takes place between maternal and foetal blood. The placenta also enables the foetus to eliminate waste products.

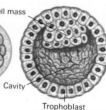

Inner cell mass

Cavity

Trophoblast

The fertilized egg cell multiplies by division to form a hollow morula *(above)* Blood supply to the placenta *(right)* and to the foetus *(left)*

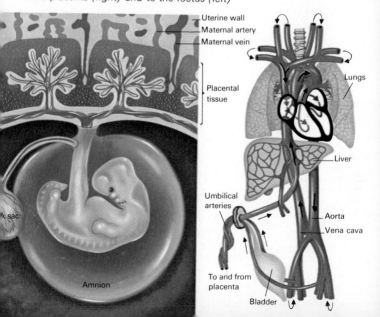

Uterine wall
Maternal artery
Maternal vein

Placental tissue

Lungs

Liver

Umbilical arteries

Aorta
Vena cava

k sac

Amnion

To and from placenta

Bladder

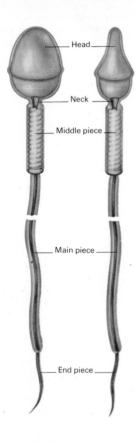

The testes

The testes are formed in the abdomen and descend into the scrotum before birth. In the adult they are $1\frac{1}{2}$ inches long and egg-shaped, and are enclosed in a tough fibrous coat. Under the microscope, they are found to be composed of many long, thin, coiled tubules separated by groups of interstitial cells. The tubules contain layers of germ cells, the innermost of which turn into spermatozoa.

The male hormone *testosterone* is the product of the interstitial cells. At puberty these are set into action by a pituitary hormone, and they continue to secrete throughout adult life. Testesterone is responsible for male secondary sex features – growth of the genitals; hair growth all over the body but especially on the pubis, face and armpits; muscle development;

Structure of the human spermatozoon (*top*) and microscopic section through a testis (*right*) showing meiotic division of semeniferous tubule cells

(*Opposite*) Longitudinal section of a testis (*left*) and through the male pelvic region (*right*)

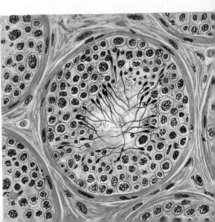

enlargement of the larynx, causing the voice to break; and sperm formation.

Spermatozoa have a head, a middle piece and a tail. The head contains the nucleus. The middle piece encloses mitochondria and is the energy source, and the 1/20 mm-long tail bends from side to side and propels the sperm forwards at up to an inch per minute.

Seminal fluid contains many million spermatozoa, but these account for only a small fraction of its volume, most of which is a nourishing medium for them, rich in fructose. The bulk comes from the epididymis (which sits on the testis), the seminal vesicles and prostate gland (at the base of the bladder), and further small glands which empty into the urethra. The testis and epididymis communicates via the vas deferens with these other glands, and in intercourse the combined products are expelled. Among the events in intercourse are erection (distension with blood) of the penis and of external female genitalia; outpouring of vaginal and other secretions to lubricate entry of the penis into the vagina; and reflex contraction of the male urethra, causing ejaculation of semen.

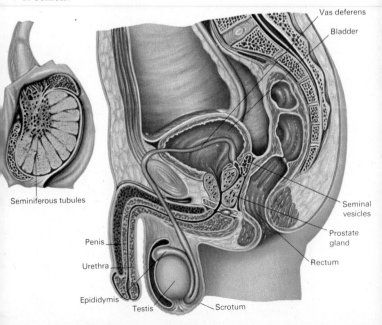

Seminiferous tubules

Vas deferens

Bladder

Penis

Urethra

Epididymis

Testis

Scrotum

Seminal vesicles

Prostate gland

Rectum

BLOOD

In the smallest organisms oxygen is available for vital processes simply by diffusing through the surface, while foodstuffs, after ingestion, can be distributed in a similar way. Beyond a certain size, diffusion is insufficient for the needs of the organism and oxygen and food transport is carried out by a specialized liquid tissue, the blood, which circulates within enclosed cavities. The chief function of the blood, the conveying of oxygen to the cells of the body, is indicated by its colour, which is due to a pigment that can pick up oxygen and release it readily. In birds and mammals the pigment is haemoglobin, which contains iron, but not all species have red blood – some invertebrates, for instance, have blue blood containing a copper compound.

Not only does the blood carry oxygen to the tissues but it takes up the waste materials of cell metabolism (such as carbon dioxide and protein breakdown products) and carries them to the lungs and kidneys, where they are eliminated. It carries foodstuffs, mineral substances and hormones to the cells, and it

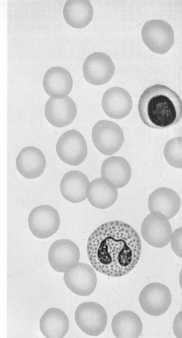

Blood smears showing two white cells (*top*) and cluster of platelets (*below*)

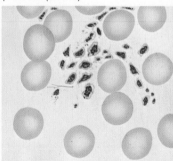

Plan and side view of a red cell

protects the tissues of the body through its anti-infective components, the white cells and the plasma antibodies. When sealed in the vessels the blood is a stable tissue, and it responds to a breach of its confines by clotting, which is due to properties of the blood itself.

Clotting demonstrates that the blood is not just a simple solution of a dye, like red ink, but a composite fluid. A clot is derived mainly from the solid elements of the blood, which make up about 45 per cent of its volume, while the serum that oozes out comes from the plasma, the liquid part. Under the microscope the solid elements are seen to be of three types – the red cells (the haemoglobin-containing, oxygen-carrying ones), the while cells (of various sorts, as shown by special stains, and involved in the response to infection) and the platelets (which take part in blood clotting). In every cubic mm of blood there are usually about 6 million red cells, between 5,000 and 10,000 white cells, and half a million platelets. The platelets are the smallest, and are only 1/400 mm in diameter: red cells are 1/140 mm across while white cells vary from 1/140 to 1/60 mm in diameter. Red cells are unusual in that they have no nucleus. They are biconcave, this shape being the most efficient for oxygen transport.

Mammals have five types of white blood cell

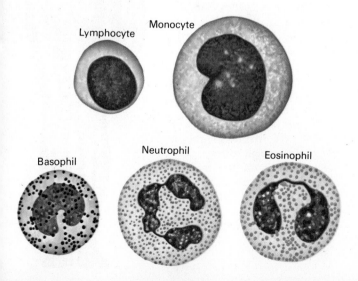

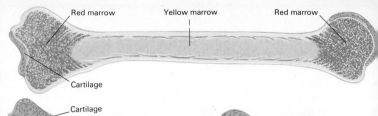

All bone is active in early life (*lower*). In adults, yellow marrow replaces red (*top*) which becomes restricted to the areas shown (*right*).

The bone marrow

In adults the red cells, platelets and most of the white cells are produced in the bone marrow. The centre of nearly all the bones is filled with this gelatinous substance, and early in life all of this marrow is red and packed with blood-forming cells. Later on much of it becomes yellow and fatty, and by adult life red marrow is restricted to the spine, breastbone, ribs, pelvis and skull. If necessary, however, blood formation can be increased by the takeover of yellow marrow in the long bones, and it is not unusual to find red tissue in the centre of the thigh bones.

Examination of the marrow, stained with dyes, shows that the mature red and white cells and platelets in the blood are at the end of a line

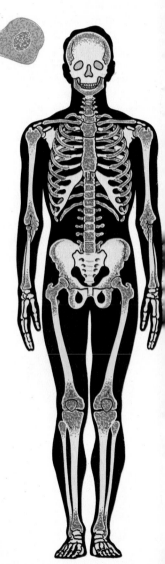

of ancestor cells. The most primitive precursor of the red corpuscle is the proerythroblast, which is up to 1/50 mm across and has a large nucleus and a blue-staining cytoplasm. This turns into the erythroblast and then the normoblast, which is smaller, has a smaller nucleus, and a pink-staining cytoplasm due to the formation of haemoglobin within it. The normoblast is the same size as a mature red cell, but differs from it in having a nucleus. This is then got rid of, producing the young red cell or reticulocyte which still has wisps of blue-staining in its cytoplasm.

The precursors of all the granular white cells and of some of the non-granular ones (lymphocytes and monocytes) are mingled in the marrow with red cell precursors. Myeloblasts turn into the smaller myelocytes, which have round nuclei and granules of various sorts, and then into white cells, whose nuclei are segmented like a string of sausages. Platelets are formed from large (1/25 mm) multinucleated cells, the megakaryocytes, which break up into little pieces. Another cell found in the marrow is the plasma cell, which makes antibodies which are released into the blood.

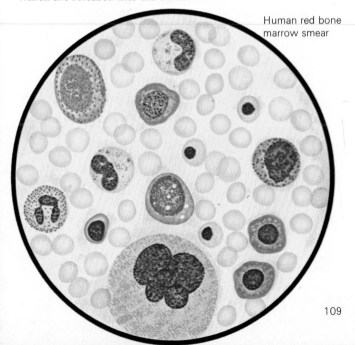

Human red bone marrow smear

109

Red cells, iron, B₁₂ and anaemia

The normal biconcave 1/140 mm red cell is filled with the iron-containing compound haemoglobin, which can combine reversibly with oxygen and carbon dioxide and is the basis of respiratory gas exchange. In 100 ml of blood there is normally about 14 grams of haemoglobin, and anaemia exists where there is a smaller quantity. Whatever the cause, anaemia can be measured by dissolving the red cells and examining the colour of the resulting solution.

Iron deficiency is the commonest cause of anaemia. In the body there is about 4·5 grams of iron, over half of which is in haemoglobin, and it is under normal conditions obtained from the diet and lost from the body in small amounts. If loss exceeds intake, haemoglobin will fall. This occurs with prolonged bleeding (as in the case of women with heavy periods) or if the diet is inadequate (as happens sometimes in old age). The anaemic blood contains fewer cells than normal and they are paler and smaller than usual.

For the maturation of cells in the bone marrow, several hormones and dietary factors are essential. These include thyroxine, vitamin C, folic acid and vitamin B₁₂. Particularly

Hypochromic anaemia in venous blood cells *(left)*.
The smear shows pale cells due to insufficient haemoglobin.
The bone marrow smear *(right)* shows the effect of haemorrhage resulting in large numbers of red cell precursors.

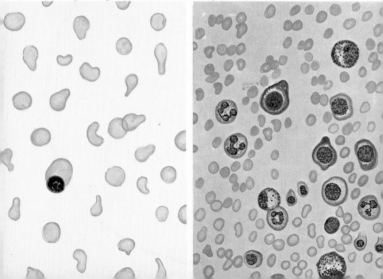

if the last two are deficient, as happens when they are inadequately absorbed from the gastrointestinal tract, this maturation is delayed and the cells which are released from the marrow into the blood may be larger than usual and nucleated. Vitamin B_{12} is a deep red, cobalt-containing substance, found in meat, eggs and milk and normally absorbed from the small intestine. Very small amounts are required for health. In *pernicious anaemia* the stomach lining is abnormal and absorption is impaired; anaemia and nervous damage result.

Anaemia also occurs when red cell production is less than red cell breakdown. These cells normally live for 3 to 4 months before being destroyed in the spleen and marrow. Destruction may be excessive in a variety of conditions in which there are abnormal forms of haemoglobin or abnormal enzymes in the cells, or in cases where the red cells are weakened by the presence of malarial parasites. In these states *haemolysis* occurs, and the released haemoglobin is converted into the bile pigment bilirubin which may enter the blood in sufficient quantity to produce obvious jaundice.

Polycythaemia is 'too many red cells'. It happens when the marrow is stimulated excessively – often without clear cause – and leads to a purple-red complexion rather than a pallid one.

Pernicious anaemia results in large, pale red blood cells and many lobed nuclei in white cells *(left)*. In bone marrow *(right)* large red cell precursors form.

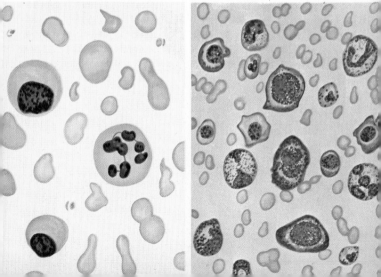

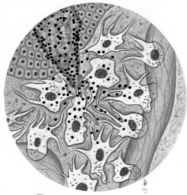

Phagocytic white cells engulf bacteria at a skin wound (*top*)

The lymphatic system of the body (*left*) and the head (*bottom*)

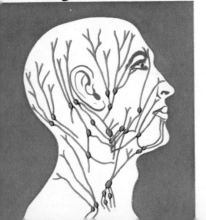

White cells and defence against infection

White cells are of two main types – those with and those without granules in their cytoplasm. The granular cells, which are formed in the marrow, are phagocytic and can engulf foreign matter (such as bacteria), and in the presence of tissue damage they are stimulated to activity. Inflammation is due to a localized increase of blood flow in the damaged area and a leakage of fluid and cells from capillary blood vessels. These are the reasons for the redness and swelling. Pus consists of many millions of granular white cells released from the blood to ingest bacteria or dead material, and its presence is the sign both of abnormal local conditions and of a normal body response.

Most of the non-granular cells are lymphocytes. These are made in the spleen and in the lymph nodes, which are found in chains all over the body but especially in close relation to the main blood vessels. The nodes contain areas of active cell production; lymphocytes pass from here into lymph channels which eventually drain into the superior vena cava and

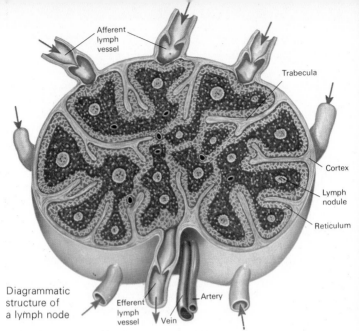

Diagrammatic structure of a lymph node

Afferent lymph vessel

Trabecula

Cortex

Lymph nodule

Reticulum

Efferent lymph vessel

Vein

Artery

the right atrium. Cells formed in the spleen go directly into the bloodstream. Besides these lymphocyte-producing tissues there are several others containing large collections of lymphocytes. These include the tonsils and adenoids.

When exposed to foreign materials, like transplanted tissues or micro-organisms, lymphocytes multiply and make antibodies in an effort to neutralize the strange substances. They are thus involved in the fight against infection and in the rejection process after spare-part surgery.

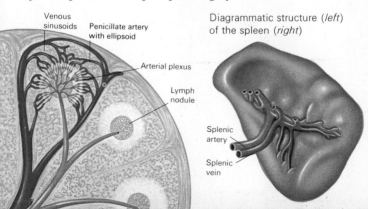

Venous sinusoids

Penicillate artery with ellipsoid

Arterial plexus

Lymph nodule

Diagrammatic structure (*left*) of the spleen (*right*)

Splenic artery

Splenic vein

Monocytes, which number a few per cent of the white cells in the blood, resemble large lymphocytes but have an indented rather than a round nucleus. They are phagocytic and are the representatives in the blood of the *reticulo-endothelial system*, which comprises similar cells in connective tissue (macrophages), liver (Kupffer cells), spleen and lymph nodes. These are all scavengers, removing damaged or dead cells and foreign particles, and then destroying or storing them.

Plasma and plasma proteins—immunization

The fluid part of the blood, the plasma, is a colourless solution containing proteins and smaller amounts of salts, glucose, aminoacids and other substances. In normal plasma there is about 7·5 grams of protein per 100 ml compared with the equivalent of less than 1 gram of sodium chloride. The plasma proteins in terms of quantity are the most important plasma

(Left) Separation of plasma proteins by electrophoresis shows the relative proportions of each component. The gamma globulins are the antibodies involved in the immune response to disease and are more in evidence in the left-hand strip. *(Right)* Proportion of cells and plasma in blood

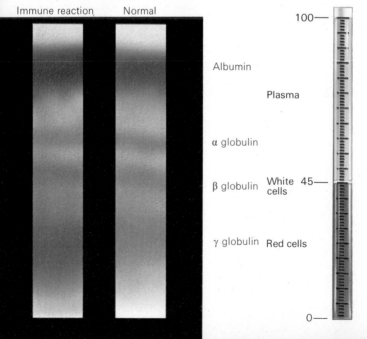

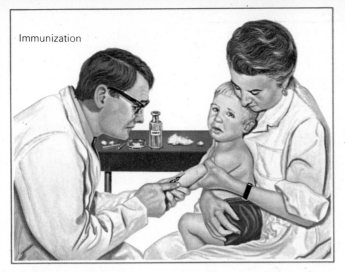

Immunization

constituents, and by their osmotic effect they help to keep fluid within the blood vessels and prevent it from accumulating in the tissues.

The different proteins can be separated in the laboratory by means of a process involving a powerful electric field (electrophoresis). About two-thirds of the protein is albumin, which is of relatively low molecular weight and is formed in large quantity in the liver. In starvation or in liver disease, much less is manufactured, and in some kidney diseases considerable amounts are lost in the urine: in these conditions the osmotic pressure of the blood falls and fluid leaks into the tissues, producing oedema. Most of the remaining third of the plasma proteins is constituted by the globulins. Three main groups are recognized – alpha, beta and gamma. They have large molecules, gamma globulin being over twice the size of albumin, and they are synthesized by the cells of the reticuloendothelial system. New gamma globulin is formed as a result of infection, for it contains the specific antibodies against infecting agents. 'Active' immunization involves antibody production in response to a small dose of a bacterium or virus. The organism need not be alive; provided its antigenic components are not damaged by killing it, a dead organism is effective in eliciting an antibody response.

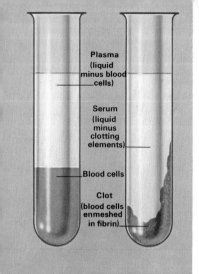

Plasma (liquid minus blood cells)

Serum (liquid minus clotting elements)

Blood cells

Clot (blood cells enmeshed in fibrin)

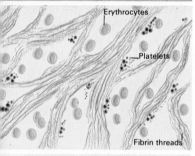

Erythrocytes

Platelets

Fibrin threads

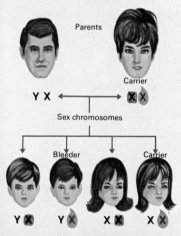

Parents

Carrier

Y X ⟷ X X

Sex chromosomes

Bleeder

Carrier

Y X   Y X   X X   X X

Blood coagulation

Fibrinogen, another plasma protein, is present in the blood only in small amounts – about 0·3 grams per 100 ml. It is a vital substance for the clotting of blood, which consists essentially of the conversion of fibrinogen to fibrin. This forms a fibrous mass and together with enmeshed platelets and red and white cells makes up a clot. The defibrinated plasma is then serum.

Platelets are very fragile. If vessel walls and tissues are damaged, as occurs whenever there is bleeding, they break up and release substances which react with a number of other substances in the blood and tissues to form thromboplastin. *Prothrombin* is yet another plasma protein, manufactured in the liver and present in the bloodstream in small quantities; in the presence of calcium

If blood is prevented from clotting, cells separate out from plasma (*top left*). On clotting, a clear fluid (serum) remains (*top right*).

Microscopic section of a blood clot (*middle*) and haemophilia inheritance (*bottom*)

Substances involved in clotting (*opposite*)

salts, it is converted by the newly-produced thromboplastin into thrombin. This is the agent that induces clot formation, for it is a protein-destroying enzyme which digests part of the fibrinogen. The remainder of the fibrinogen then combines with itself to form fibrin.

It follows that a number of different conditions can produce abnormal bleeding. If the capillaries are excessively fragile, as occurs in vitamin C deficiency, tiny spots of blood appear in the skin (purpura). If, in contrast, one of the major factors needed for clotting is deficient, large scale bleeding may result. Thus in haemophilia one of the plasma components necessary for the formation of thromboplastin is present in much reduced quantities, and so small cuts and bruises bleed uncontrollably. This disease occurs almost entirely in males; it is an hereditary condition transmitted on an abnormal X chromosome by females, in whom it is hidden due to the presence of the other, normal, X. A similar but milder disorder called Christmas disease is inherited in a different way. Excessive bleeding also happens when the platelets are deficient in number, as in severe disease of the bone marrow; in biliary disease and vitamin K deficiency, which produce a reduction in prothrombin levels in the blood; and in certain rare conditions in which fibrinogen is absent.

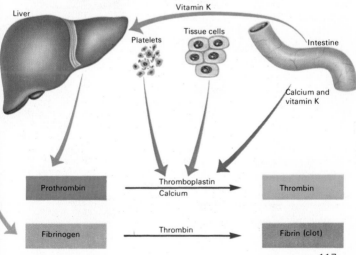

Blood groups and transfusion

Before the discovery of blood groups at the turn of the century, blood transfusion was a hit or miss procedure, sometimes successful but sometimes totally disastrous. We now know that immune reactions are responsible for 'incompatible' blood transfusions, for red cells possess certain antigens and plasma contains related antibodies. The presence of these is determined genetically; antibody is never present in the same blood as its related antigen, for if the two meet the red cells mass together and then dissolve. This, however, is precisely what happens when incompatible blood is transfused – disaster may follow, because the damage to the red cells can lead to fatal shock and kidney damage.

There are two antigenic red cell substances, known as A and B. They are complex sugars. Red cells can be of four kinds – A, B, AB (containing both) and O (containing neither). In Great Britain O and A are common (46 and 42 per cent) while B (9 per cent) and AB (3 per cent) are rare. The serum of someone belonging to group O contains antibodies to A

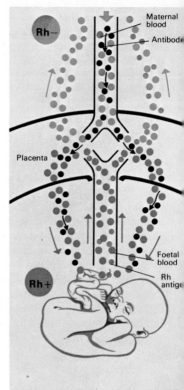

Foetal blood is incompatible with maternal blood in a Rhesus baby (*above*)

The effect of serum A and B on blood cells determines blood groups. Tests are carried out on glass slides (*opposite*). Serum A, containing anti-B, agglutinates both B and AB cells. Transfusion of B or AB cells would therefore stimulate an anti-B reaction in a group A recipient; O cells, with no antigens, would not be reacted to.

and B, while a person in group AB has neither and someone in A or B has antibody only to the other substances. Before transfusion can be carried out, the cells of the donor and the recipient must be grouped against serum of known groups while in addition the cells to be transfused must be tested directly with the recipient's serum. This is the only real test of compatibility, for apart from the ABO system of groups there are many other groups, mostly of minor importance but occasionally responsible for severe reactions.

The other major blood group system is the Rhesus system. The Rhesus 'factor', which is a collection of antigens, was first found in the red cells of Rhesus monkeys, but it is present in 85 per cent of humans (who are termed Rh+) and absent in 15 per cent (Rh−). Rhesus *antibodies* are not usually present in the plasma of this 15 per cent. Like other blood group substances, the 'factor' is inherited, and its practical importance lies in the observation that the child of a Rh+ father and Rh− mother may itself be Rh+. If this occurs, the Rh+ cells of the foetus will set up the formation of antibodies in the mother's circulation. These antibodies will then enter the blood of the foetus and destroy its red cells in the way already described. A Rhesus baby – severely jaundiced, anaemic, premature and sometimes spastic – is the result. Its bone marrow makes up for red cell loss by releasing erythroblasts, but their quantity is inadequate and often the only way to save the baby is by transfusing healthy blood to replace the abnormal blood.

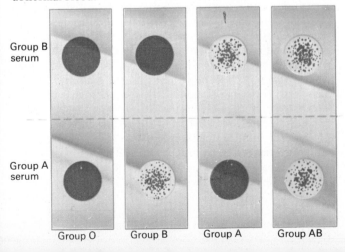

Group B serum

Group A serum

Group O   Group B   Group A   Group AB

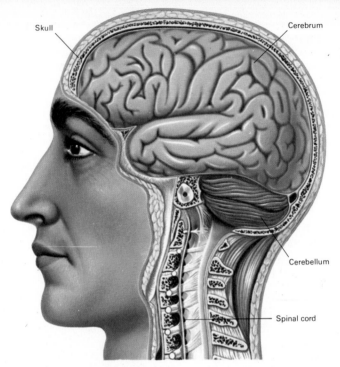

Skull

Cerebrum

Cerebellum

Spinal cord

THE NERVOUS SYSTEM

The nervous system is made up of the brain, spinal cord and nerves. It has two basic functions – the detection and processing of information from within and outside the body, and the production and control of movement, which is brought about by muscle action. The brain is also the seat of consciousness, and the mind is in some way a reflection of its activity.

The adult brain weighs about 3 pounds. It is divided into the *cerebrum,* which is the largest part, the *brainstem,* which joins up with the spinal cord, and the *cerebellum,* which sits behind the brainstem below the cerebrum and is concerned with the fine control of movement. The surface of the cerebrum is wrinkled with convolutions separated by fissures; this wrinkling creates a large surface area and allows many brain cells to be packed into a limited space. Groups of convolutions form several distinguishable lobes, each of which has a particular function. When the cerebrum is sliced, it is seen to consist of an outer grey layer, the cerebral cortex, beneath which is a mass of white tissue. Within the brain are four interlinked cavities or *ventricles,* through which fluid circulates, acting as a shock absorber for the delicate tissue. Around the ventricles in the cerebrum are several masses of grey matter known as the basal ganglia. They are grey and translucent like the cortex because they are full of nerve cells; the white matter appears pale and opaque in contrast because it is made up of nerve fibres with fatty insulating sheaths.

The uppermost part of the brainstem, the *midbrain,* is tucked up into the cerebrum. It contains a number of movement-coordinating centres as well as being responsible for sleep and wakefulness. Beneath this is the *pons* (bridge) connecting with the cerebellum, and joining the pons to the spinal cord is the *medulla,* in which are groups of nerve cells regulating such vital functions as respiration and blood pressure. The surface of the cerebellum is finely folded into *folia* (leaves), which resemble the cerebral convolutions in having outer

(*Opposite*) The brain enclosed in the cranial cavity

Areas of the body supplied by the spinal nerves. The orange area is supplied from the cervical segments of the spine, the yellow area from the thoracic segments, the green area from the lumbar segments, and the pink area from the sacral segments.

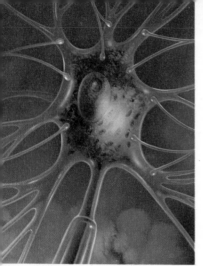

(*Above*) Cell body of a motor neuron
(*Below left*) A nerve impulse crosses a synapse by causing release of a chemical transmitter stored in the synaptic vesicles. The transmitter generates a new impulse on the other side of the cleft.
(*Below right*) Diagram of motor neuron showing route of nerve impulse.

grey and inner white layers.

The spinal cord runs from the base of the skull to the level of the upper lumbar spine. While the brain itself gives off twelve pairs of nerves, serving sensation and movement in the head and neck, the cord gives off thirty-one pairs supplying the trunk and limbs. An H-shaped core of grey nerve cells forms a column in the centre of the cord, surrounded by white nerve fibres. The front prongs of the H contain cells responsible for movement, while the back prongs are involved in sensation. The white matter contains ascending sensory pathways to the brain, and descending motor pathways. Each spinal nerve is made of two nerve roots, motor and sensory, which emerge from the front and back of the cord and join together.

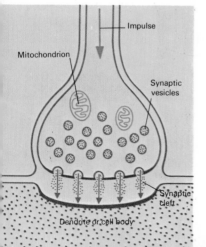

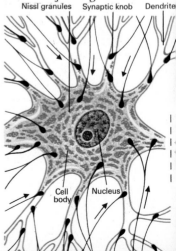

Neurons and nerve fibres

The structural and functional unit of the nervous system is the nerve cell or *neuron*; nervous tissue is composed of neurons, their elongated outgrowths, and a form of connective tissue known as *neuroglia*. Neurons are electrically charged cells which, once formed, cannot be renewed or replaced. They give off long extensions which constitute nerve fibres and may be three feet or more in length. These specialized structures, in continuity with the body of the cell, conduct nerve impulses (which are electrical waves) at high speed and in so doing carry messages within, to, and from the brain.

Neurons in different parts of the brain and cord vary widely in size and shape. The granule cells in the cortex of the cerebellum are only 1/200 mm in diameter while the big motor cells in the spinal cord reach 1/8 mm across. However, virtually all types are composed of a cell body, an *axon,* which arises from the axon hillock of the cell and may run a considerable distance as a nerve fibre, and *dendrites,* which are finely branching protoplasmic processes.

The cell body can be stained to show Nissl granules, which disappear when the cell is damaged or is repairing itself. Axons are fine cylinders which convey information to or from the cell body and, in the brain, end by branching to meet the dendrites of other nerve cells at junctions called synapses. Learning, memory, and information storage within the brain involve not only biochemical changes in the nerve cells but also the forming of new synapses. Dendrites (absent in some

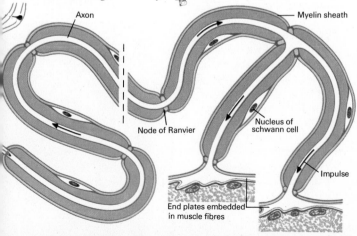

Axon

Myelin sheath

Node of Ranvier

Nucleus of schwann cell

Impulse

End plates embedded in muscle fibres

specialized receptor cells concerned with vision, balance and hearing) may connect the parent cell with the axons of many other neurons and thus act as a vast receptor area for messages.

Neuroglia and Schwann cells

The neuroglia, the supporting elements, are of three main types – astrocytes, oligodendrocytes and microglia. They fill up the space between the neuronal structures and are important in the maintenance of nerve function. *Astrocytes* are the largest, and are star-shaped cells with branching processes. They are found chiefly in the grey matter and are involved in the nutrition of neurons and the repair of brain damage.

Oligidendrocytes are smaller and have fewer branches. In white matter they lie singly or in rows between nerve fibres, the fatty insulated coverings of which they produce. *Microglia* are the smallest of the glial cells and are scattered throughout the brain. They are the scavengers; they may pick up remnants of damaged cells and move around.

Two special types of neuroglia occur in the ventricles

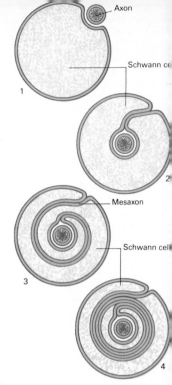

The progressive development of a myelin sheath is illustrated above. The drawing below shows a couple of myelinated nerves.

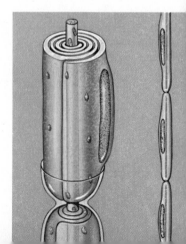

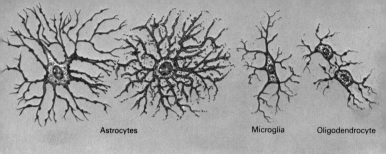

The three types of neuroglia cell

of the brain. The ependyma is a layer of columnar cells with a ciliated border covering their inner walls, the aqueduct joining the third and fourth ventricles, and the central canal of the cord. The choroid plexuses, which produce cerebrospinal fluid, are gland-like folded masses of similar, flatter cells.

The nerve fibres which exit from the brain and cord make up the nerves of the head, trunk and limbs. They are of two sorts – those with and those without a fatty covering known as the *myelin sheath*. In peripheral nerves myelin is formed by *Schwann cells*, which envelop the nerve axons and cover them with layer upon layer of protein and fat. This myelin sheath acts as an electrical insulator and enables nerve messages to be conducted faster; the thicker axons, which are anyway faster than thinner ones, are myelinated, while small-diameter axons are without such a sheath. Myelin is laid down in segments, which may be several millimetres long. Each segment is produced by one Schwann cell, and the junction between two is at the structure called the *node of Ranvier*.

The detailed structure of a node of Ranvier

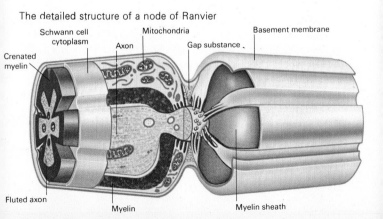

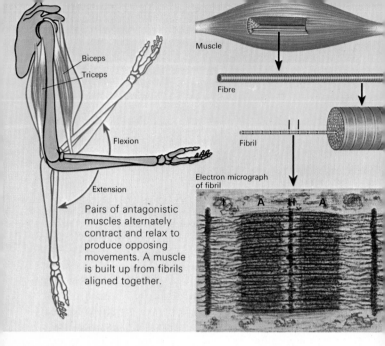

Biceps

Triceps

Flexion

Extension

Pairs of antagonistic muscles alternately contract and relax to produce opposing movements. A muscle is built up from fibrils aligned together.

Muscle

Fibre

Fibril

Electron micrograph of fibril

Skeletal muscle

The muscles of the head, trunk and limbs are known as voluntary muscles, for they contract and produce movements in response to conscious efforts of will. They are also called striped muscles, for under the microscope their long thin fibres are seen to have fine dark and light cross markings. These are the A band and the I band, and they contain two proteins, respectively myosin and actin. The two form a complex substance, *actomyosin;* its molecules are able to contract in response to nervous messages. This is the basis of movement. Actomyosin can be extracted from muscles and under special conditions converted into threads which shorten in the presence of the compound ATP, providing a demonstration model of muscular contraction. When muscles shorten, energy is used up and heat is liberated. Energy comes from glucose, itself derived from glycogen stores in muscle, and lactic acid is formed. This accumulates in muscles which are used heavily for any length of time or deprived of adequate blood flow, and an excess amount causes cramp.

Muscles make up the bulk of soft tissue in a well nourished person, and are familiar as red meat. How is it that an infinite variety of movements can be produced by structures that are anchored at both ends and can merely contract and relax? The answer lies in the complexity of the muscle arrangement in the body and the fact that for every muscle there is another which has the opposite action – its antagonist. This enables every movement to be controlled in force and range, while the intricate arrangements of muscles across joints allow the greatest possible diversity of action. An example of antagonism is provided by the biceps and triceps. The biceps is attached by tendons above to the shoulder joint and the humerus and below to the radius. When it contracts the forearm is brought closer to the shoulder. The triceps runs from the top of the humerus across the elbow joint to the upper end of the ulna, and its contraction therefore straightens the elbow and the whole arm. When the biceps is in action the triceps is relaxed, and vice versa.

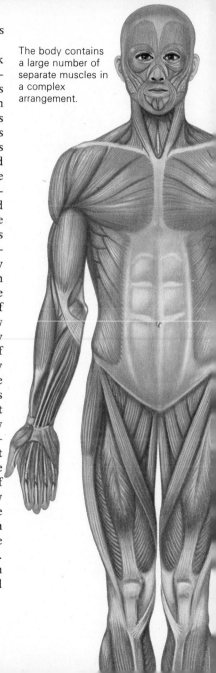

The body contains a large number of separate muscles in a complex arrangement.

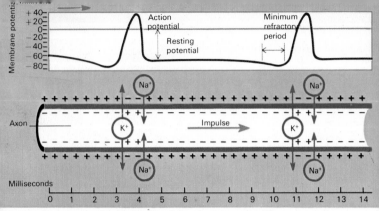

Changes in electrical potential and movement of ions across the nerve membrane during an action potential

Voluntary movement

Willed movements begin as electrical activity in large neurons of the cerebral cortex. Nerve impulses send messages from these cells along myelin-enclosed nerve fibres to the motor neurons of the spinal cord. The axons of these cells constitute motor nerve fibres which convey impulses to muscle. The nerve messages arrive at special neuromuscular junctions and induce contraction of the muscle fibres.

The changes that occur in nerves and muscles during such activity are both electrical and chemical in nature. In the resting state, there are many potassium ions and few sodium ions inside the nerve cell and axon, and little potassium and much sodium outside. These concentration differences are due to properties of the thin cell membrane surrounding each neuron and its nerve fibre. This membrane is also responsible for the electrical charge of the nerve cell (about 60 millivolts, the inside of the cell being negative in relation to the outside). Nerve cell activity and nerve impulses consist basically of very brief changes of electrical charge and ionic concentration. The

Local current flow during passage of action potential.

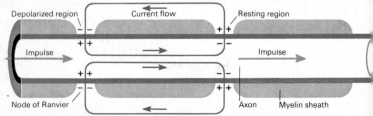

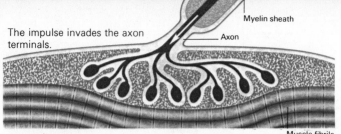

The impulse invades the axon terminals.

Myelin sheath

Axon

Muscle fibrils

electrical change is the *action potential,* a spike of positive voltage lasting a few thousandths of a second, and this is associated with a brief reversal of the normal distribution of sodium and potassium. The nerve impulse is formed by the passage of the action potential along the axon at speeds up to 200 feet per second. In the faster conducting myelinated fibres these electrical and chemical events are confined to the nodes of Ranvier, enabling a very rapid stepwise transmission of the impulse to occur.

On arriving at the neuromuscular junction, the nerve impulse releases from the nerve endings a minute quantity of the compound acetylcholine, which passes across to the muscle, becomes attached to special sites and causes rapid changes in electrical charge and ionic concentrations like those in conducting nerve tissue. The changes continue throughout the muscle fibres and initiate contraction. Acetylcholine lasts only for a short while, and is rapidly destroyed by the enzyme cholinesterase. Curare (the South American blowpipe arrow poison) paralyses by becoming attached at the neuromuscular junction and so blocking the normal action of acetylcholine.

A tracing of the tension developed in a single muscle twitch

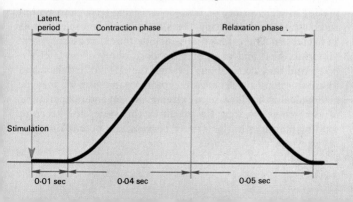

Latent period

Contraction phase

Relaxation phase .

Stimulation

0·01 sec   0·04 sec   0·05 sec

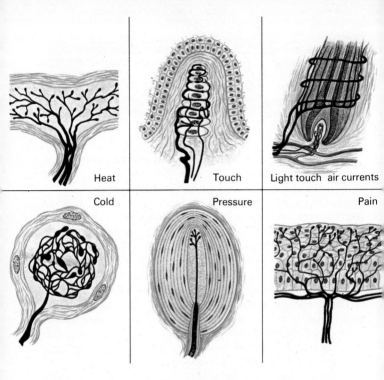

Heat   Touch   Light touch air currents

Cold   Pressure   Pain

The receptor nerve endings in the skin subserving different
sensations have different structures although they are all basically
dendrites of sensory neurons.

Sensation and sensory receptors

In addition to the special senses – sight, hearing, taste and
smell – the body possesses the means of detecting and inter-
preting a wide variety of sensations arising from both within
and without and affecting the skin, viscera, muscles, joints
and tendons. These sensations include touch, pressure, pain,
vibration, heat and cold, and the positioning of parts of the
body, and they come about through the effect of external and
internal stimuli on sensory receptor organs. Stimuli may be
mechanical or (in the case of extremes of pain and temperature)
physico-chemical, but the result is the same, the receptor
sending messages in the form of nerve action potentials to the

central nervous system. Sensory receptors are thus *transducers* and can be compared with microphones or record player cartridges, which convert a mechanical input into an electrical output.

The sensory receptors of the body show widely varied structure under the microscope. Structure is related to function, and the simplest receptors, which are merely nerve endings in the skin, are responsible for detecting pain, while complex basket-like or onion-shaped receptors respond to light touch, small movements and pressure. All however are

Impulses from receptors in the skin pass through the spinal cord to the brain. Some axons pass right up the cord before synapsing in the medulla; others synapse upon entry into the spinal cord.

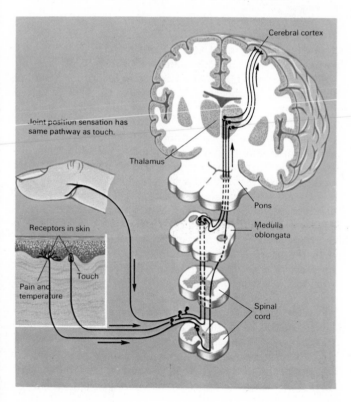

Cerebral cortex

Joint position sensation has same pathway as touch.

Thalamus

Pons

Receptors in skin

Medulla oblongata

Touch

Pain and temperature

Spinal cord

131

the endings of sensory nerves, and the information they send is in the form of nerve impulses. As the nerve action potential is constant in size and duration, change in input is signalled to the brain as a change in the frequency of impulses.

Sensory messages in sensory nerves enter the central nervous system in the posterior roots of the spinal cord. These nerves are the extensions of nerve cells lying in a small swelling (ganglion) of the posterior root, and the centrally-travelling processes of the same cells carry the messages into the cord, through which they eventually reach the cerebral cortex. In the cord the first sensory fibres may transmit their information to other cells, and between the cord and the cerebrum are relay stations, which analyse the messages received.

Reflex activity—
the organization of the spinal cord

Reflexes are unconscious actions in response to sensory stimuli. The withdrawal of a hand from something very hot that has been touched inadvertently, and the kick produced by a shap tap on the tendon just below the kneecap, are both reflex actions. These and all the many other reflexes are mechanisms of the central nervous system, having in common an incoming sensory and outgoing motor pathway. They are inborn and involuntary actions, and the spinal cord and brainstem are responsible for most of them, so that they may still be present even after a great deal of brain damage has occurred. 'Conditioned reflexes', which are acquired responses based on these actions, are rather different for they involve the cerebral cortex and cannot be established in its absence.

In the simplest reflex, the knee jerk, the tap below the kneecap briefly stretches the tendon, stimulating stretch receptors in it which send off a volley of nerve impulses to the spinal cord. The incoming message is carried directly to the motor neuron in the anterior horn of the cord at the same level, and this neuron then fires off a set of impulses which travel along its axon and reach the muscle at the front of the thigh, which contracts, straightening the leg and thus combating the original stretch to the muscle and its tendon. The reflex thus has a *protective* function, preventing damage to the muscle, and all other reflexes either prevent injury or help to restore the normal state of the body.

The knee jerk must in fact be far more complex than the simple description just given, because for maximum efficiency contraction of a muscle is accompanied by relaxation of its antagonist. In this case the hamstring muscles at the back of the thigh relax while the others contract, and this relaxation is produced reflexly by a suppression of the activity of the hamstring motor nerve cells. This suppression occurs through the agency of intermediate neurons in the cord, for the incoming sensory impulses from the stretched tendon are also relayed up the spinal cord along a number of pathways.

In a spinal reflex, information from receptors is channelled to motor neurons innervating appropriate muscles without passing through centres in the brain. With the knee jerk there is only one synapse between the sensory neuron and the motor neuron. Usually there are intervening neurons to relay the information.

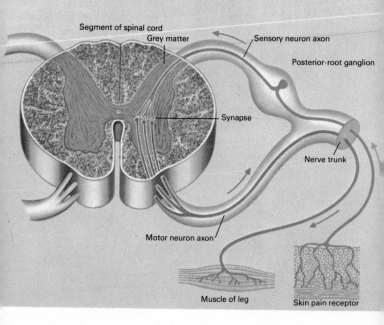

Segment of spinal cord
Grey matter
Sensory neuron axon
Posterior-root ganglion
Synapse
Nerve trunk
Motor neuron axon
Muscle of leg
Skin pain receptor

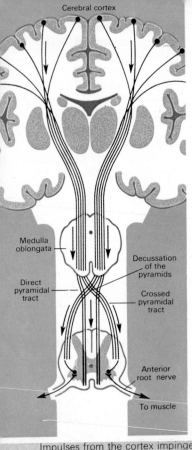

Cerebral cortex

Medulla
oblongata

Direct
pyramidal
tract

Decussation
of the
pyramids

Crossed
pyramidal
tract

Anterior
root nerve

To muscle

The control of movement

Normally muscular move-
ments are smooth and co-
ordinated because of reflex
activity, nervous feedback
(servo) systems, and the mass
of ever-changing information
impinging on the spinal motor
neurons. If the normal working of the brain and spinal
cord is disturbed, movements
may become weak, stiff or
jerky and complicated actions
like walking may become
impossible.

Movements begin as elec-
trical activity of the pyramidal cells of the motor area
of the cerebral cortex, which
lies just in front of the central
furrow. From these cells the
fibres pass downwards and
are grouped together in the
internal capsule deep in the
brain. They then descend in
front of the midbrain as the

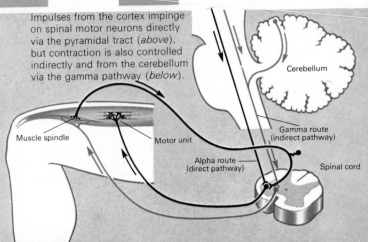

Impulses from the cortex impinge
on spinal motor neurons directly
via the pyramidal tract (*above*),
but contraction is also controlled
indirectly and from the cerebellum
via the gamma pathway (*below*).

Cerebellum

Muscle spindle

Motor unit

Gamma route
(indirect pathway)

Alpha route
(direct pathway)

Spinal cord

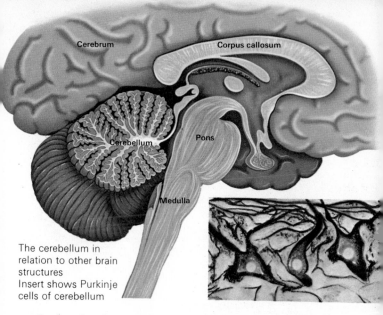

The cerebellum in
relation to other brain
structures
Insert shows Purkinje
cells of cerebellum

Labels: Cerebrum, Corpus callosum, Cerebellum, Pons, Medulla

cerebral peduncles. Lower, in the medulla, the two groups of fibres cross each other and they then go down into the spinal cord and end in contact with the motor neurons. This crossing over explains why in strokes the damage is in one half of the brain while the paralysis is on the other side of the body.

The corticospinal motor fibres carry messages which set off voluntary movements by activating spinal motor neurons. These cells are influenced also by impulses from the basal ganglia, midbrain reticular formation, pons and cerebellum, which regulate their activity; they are moreover affected reflexly by the muscles themselves. Muscles are in a state of tension ('tone') due to occasional contraction of groups of their fibres. Tone is determined by a small number of special fibres which are made to contract the desired amount by impulses from small ('gamma') motor neurons. Tension receptors around the fibes then fire off and their messages, arriving at the spinal cord, activate the larger ('alpha') motor neurons which supply the rest of the muscle. When the tension in this as a whole is the same as that in the special fibres, motor neuron activity stops for the time being.

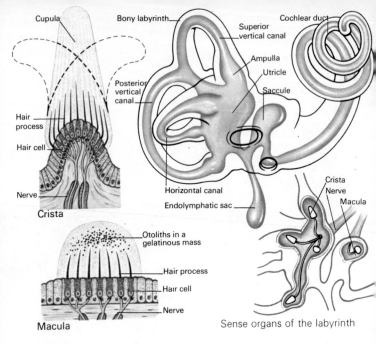

Labels on the figure:

Cupula
Bony labyrinth
Superior vertical canal
Cochlear duct
Ampulla
Utricle
Posterior vertical canal
Saccule
Hair process
Hair cell
Nerve
Crista
Horizontal canal
Endolymphatic sac
Otoliths in a gelatinous mass
Hair process
Hair cell
Nerve
Macula
Crista
Nerve
Macula

Sense organs of the labyrinth

Balance, the cerebellum and the labyrinths

The cerebellum is important in the regulation of movement and in maintaining balance, in which it works in close connection with the balance organs or labyrinths. Each cerebellar hemisphere is linked to a labyrinth, to the opposite half of the cerebrum, and to the spinal cord on the same side. This spinal connection means that damage to the cerebellum causes disturbances of movement such as clumsiness and tremors on the same side of the body as the damage (in conrast to cerebral damage). Fibres from the cerebellum also activate the 'gamma' motor neurons of the cord, and so in cerebellar disease muscular tone is diminished and the muscles may become very lax.

The labyrinths, one on each side, are situated in the inner ear, deep in the sides of the skull. Through their linkages with the cerebellum, brainstem and cerebral cortex they continually influence the motor activity of the spinal cord. They pick up information about the position of the head and analyse it so that messages can be sent to the muscles of the body to help preserve balance.

The labyrinth or vestibular apparatus is an intricate structure composed of a gravity-sensitive part (the utricle and saccule) and three semicircular canals sensitive to rotation of the head. These canals curve upwards, sideways and backwards, and are filled with fluid. In each there is a slightly expanded region, the ampulla, containing a bundle of sensitive nerves projecting into the canal. Rotation of the head swirls the fluid in the canals and pushes these nerves sideways. They then fire off nerve impulses which pass along the vestibular nerve to their cell bodies grouped together in the pons. As there are three canals at right angles to each other rotary movement in any direction can be detected.

Inside both the utricle and the saccule there is a small ridge, the macula. This is made of sensory cells with stiff, calcium-laden projections which rise vertically when the head is upright. Change in posture bends these projections; the cells sense the change, which is signalled to the brain.

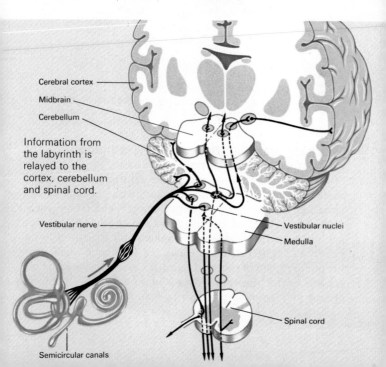

Cerebral cortex

Midbrain

Cerebellum

Information from the labyrinth is relayed to the cortex, cerebellum and spinal cord.

Vestibular nerve

Vestibular nuclei

Medulla

Spinal cord

Semicircular canals

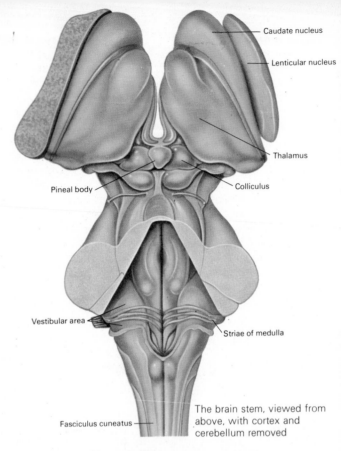

Caudate nucleus

Lenticular nucleus

Thalamus

Colliculus

Pineal body

Vestibular area

Striae of medulla

Fasciculus cuneatus

The brain stem, viewed from above, with cortex and cerebellum removed

The basal ganglia, thalamus, hypothalamus and reticular formation

These groups of neurons around the ventricles in the centre of the brain are involved not only with movement and sensation but with thirst, appetite, emotion and temperature.

The basal ganglia, comprising the caudate nucleus, putamen and globus pallidus, are closely linked with structures in the nearby midbrain (the red nucleus and substantia nigra) with the motor region of the cerebral cortex and – through the red nucleus – with the spinal cord. Collectively these neuronal groups play an important part in voluntary movement, and when they are diseased the muscles may become slow, weak and rigid. Parkinson's disease, with limited, slow and tremulous movements, results from damage to this system, and it can be treated by an operation in which some of the inner connections of the basal ganglia are severed.

The thalamus is situated near the midline and behind the basal ganglia on each side of the third ventricle. It is an egg-shaped mass of nerve cells which receives all the sensory information from the head, limbs and trunk, processes it and transmits it to the sensory area of the cerebral cortex. When the thalamus is damaged, patients lose the experience of light touch and the ability precisely to localize sensations and may feel a severe, burning pain.

Below the thalamus are a few small collections of cells, forming the hypothalamus. It regulates heat production and loss by stimulating shivering and sweating. It contains receptors which are sensitive to changes of salt concentration in the blood and control water intake and output through thirst and ADH secretion. Hypothalamic injury may cause emotional outbursts, great overeating with obesity and reduction of sexual activity, so this part of the brain is obviously implicated in the most vital bodily functions.

The reticular formation is a poorly defined network of nerve cells and fibres in the centre of the brainstem and passing down from the thalamus to the medulla. It sends axons down the spinal cord to motor neurons and it plays some part in controlling reflexes and muscle tone. In addition it is concerned with sleep and consciousness, for changes in its electrical activity produce changes of wakefulness.

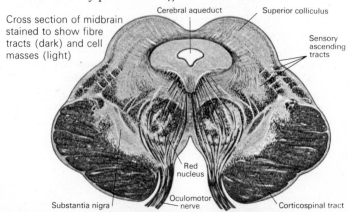

Cross section of midbrain stained to show fibre tracts (dark) and cell masses (light)

Cerebral aqueduct

Superior colliculus

Sensory ascending tracts

Red nucleus

Substantia nigra

Oculomotor nerve

Corticospinal tract

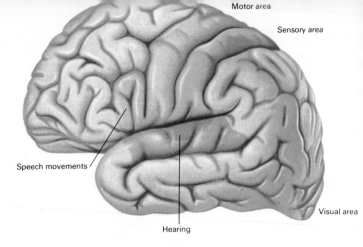

Motor area

Sensory area

Speech movements

Visual area

Hearing

Left cerebral hemisphere with main functional areas

The cerebral cortex—localization of function

The cerebral hemispheres make up the greater part of the brain and are bigger in man than in any other animal. Man's ascendency over his environment and other species is related to the size of his cerebrum. The cortex, or external layer of grey matter is crumpled and wrinkled; if stretched out flat it would cover nearly $2\frac{1}{2}$ square feet. It is on the average only 1/8 to 1/10 inch thick, but this thin layer nevertheless contains millions of neurons that are the basis of mind.

As mentioned earlier, groups of convolutions or gyri form separate lobes. The frontal lobes, placed above the orbits, are concerned with personality, behaviour and movement. The lowest left frontal gyrus is the centre for speech. Further back are the parietal areas, which are involved in the analysis of sensations and with the 'body image', this is, recognition of the body in relation to its surroundings. At the back of the head, on top of the cerebellum, are the occipital lobes, whose function is visual. On either side of the brain lies a separate projection, like the thumb of a boxer's glove. This is the temporal lobe, which interprets auditory information. A curled structure, the hippocampus, is hidden inside the temporal lobe and is partly concerned with taste and smell.

In most regions of the cortex six separate cell layers

can be recognized under the microscope. The outermost (layer 1 in illustration) contains only axons, dendrites and neuroglia. Next come two pairs of layers (2 and 3, 4 and 5) containing small round neurons and large pyramidal neurons, and an inner layer (6) of long spindly cells lying at right angles to the surface. The relative thickness of the layers varies from place to place in the cerebrum, and in certain regions special cell types are found. One of these cell types is the 'giant' pyramidal *Betz cell* of the motor region of the frontal cortex.

The motor area of one hemisphere controls voluntary movements of the other side of the body. Since information about individual movements rather than individual muscles is sent out, and since the range and variety of movements is greater for some parts of the body than for others, the cortex is not equally apportioned to all parts: the more delicate the movements, the larger the cortical area controlling them. Thus the fingers and face are governed by relatively many Betz cells, while the neck and back are controlled by relatively few.

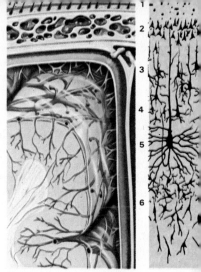

The cortex is protected by the skull, and cushioned by fluid in the cranial membranes *(left)*. On the right is a section of the cortex with some cell bodies stained — the layers are numbered as in the text. A Betz cell can be seen in the centre.

(Below) The relative representation of the body surface on the cortex

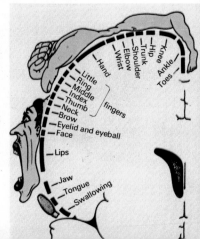

Personality and the prefrontal lobes

One particular part of the cerebral cortex that is very well developed in man is the prefrontal region, which lies in front of the motor area. It has no primary sensory or motor functions, but is connected with other cortical regions and with some of the deeper structures of the brain, including the thalamus and hypothalamus: it is involved in the control of emotions and the storage of information, and it is the part of the brain concerned above all with personality and intellect.

Damage to the prefrontal lobes may occur in disease or be produced by accident or by the surgeon in the treatment of some mental disorders. Operations such as leucotomy and lobotomy cause characteristic behaviour changes by interfering with normal function. Judgment is adversely affected and insight is lost, so that self-confidence often rises while ability decreases. A marked feature is loss of concentration; people who have had these operations cannot stick to one task for long and go on to another. In addition, emotional responses are altered. Emotions are often blunted, but they are sometimes exaggerated, with the result that an everyday situation may produce violent rages or incongruous cheerfulness. These disturbances occur not only in localized damage to the pre-

Major association pathways of the cortex

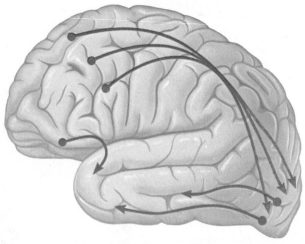

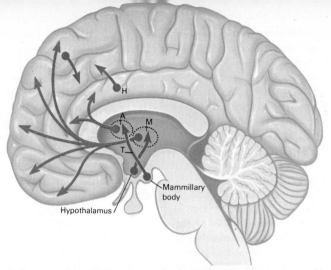

Pathways involved in emotion and memory link the hippocampus, *H*, the prefrontal lobes, the thalamus and hypothalamus and the mamillary body. *A* is the anterior nucleus of the thalamus, *M* the medial nucleus, and *T* is the mamillo-thalamic tract.

The lobes of the cerebral cortex

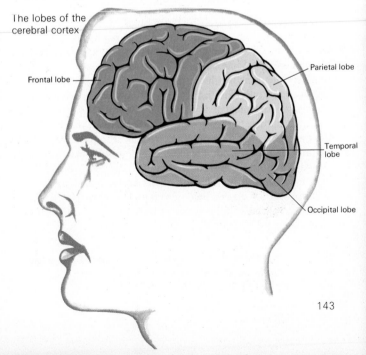

Frontal lobe

Parietal lobe

Temporal lobe

Occipital lobe

frontal lobes but also in any condition which disturbs the functioning of the cerebral cortex as a whole. The picture of dementia starts with defects of reasoning and judgment and continues with deterioration of social habits – such as language, dress and cleanliness – and emotional instability and impulsive behaviour. If the disorder is severe or progressive, the general condition of the body deteriorates as well.

Perception—the parietal lobes

Just as the frontal cortex is the highest control centre for voluntary movement, so the parietal lobes are implicated in the most precise aspects of sensation. The sub-station for sensory messages, as has been noted, is the thalamus, and some 'crude' sensations, such as pain and heat, are felt largely at this level. However, the parietal cortex is essential for the recognition of shape, size, surface texture and the various other qualities that depend on fine discrimination. It also refines the cruder sensations and so helps the body to distinguish different degrees of heat and cold, and it enables sensations to be localized and small amounts of joint movement to be assessed.

Damage to this part of the brain causes a sensory loss which is not intense but which is very disturbing and disabling for the patient: by severing the connections with other regions of the cerebral cortex it may produce much stranger effects than simple sensory loss, for the recognition of an object involves not only the pathways leading to the parietal cortex but also those concerned with memory and with producing the correct response to the incoming information. If the memory or the correct response is missing, what is seen (or heard of felt) will not be recognized. This loss of recognition may extend to the patient himself, and someone with a damaged right parietal lobe may ignore the left side of his body, believing that his left arm and leg do not belong to him, and also the left side of space, so that he will be aware of only the figures 12 to 6 on a clock face.

The illustration shows the loss of vision and awareness of left limbs resulting from a loss of information from the right cortex such as might occur in damage to the right parietal lobe.

Speech and its control

Human beings communicate with each other by uttering sounds that are shaped into meaningful patterns. Such communication is the means of expressing feelings and thoughts and conveying information and commands. This is speech, and it has three main components – the sounds themselves, their shaping or articulation, and their control or ordering.

The sounds, the first basic element of speech, are produced in the larynx by the vibration of the vocal cords in the airstream coming from the lungs. The tension of the vocal cords, and so the pitch of the sounds, is altered by the small muscles of the larynx, which are activated by the laryngeal nerves; if these nerves are affected in any way, sound production is disturbed and hoarseness may result.

Articulation depends on the muscles of the jaws, lips, tongue and palate, which both alter the shape of the resonating cavities and interrupt the stream of sound. The nerves supplying these muscles have their origin in the brainstem, but like all voluntary motor activities articulation is under the

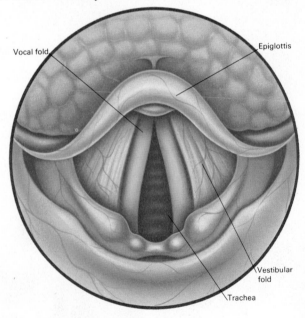

Vocal fold

Epiglottis

Vestibular fold

Trachea

control of the cerebral cortex.

Full control of the muscles of speech ensures that spoken messages are carefully and precisely uttered, but not that they are meaningful. For speech to have a meaning, rather than to consist of random syllables, the whole of one cerebral hemisphere must be working normally. In most people it is the left hemisphere which is vital for speech, but in a few left-handers it is the right one. A localized area in the left inferior frontal gyrus of the cortex is often regarded as the speech control centre, for when it is damaged (as happens in strokes) speech is severly impaired. In fact, since the various cortical regions are intricately inter-connected, most of the hemisphere is involved in normal speaking and damage anywhere in it (though particularly in the back half) can cause disturbances of speech. The impossibility of teaching chimpanzees to use more than a very few speech sounds relates to a lack of cerebral organization.

Larynx viewed from above *(opposite)*. Dissected rear view of larynx *(above)*. Position of tongue for producing consonants G and L *(right)*.

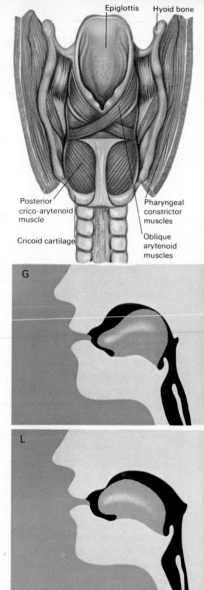

Epiglottis   Hyoid bone

Posterior crico-arytenoid muscle

Cricoid cartilage

Pharyngeal constrictor muscles

Oblique arytenoid muscles

G

L

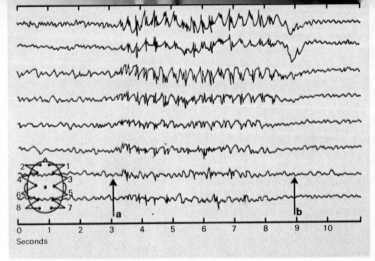

EEG record during a brief epileptic fit — note the spiky waves during the attack (starting at arrow marked *a*) and the return of the alpha rhythm after eye closure (arrow marked *b*). The changes are clearest in the records from the back of the head.

The EEG and epilepsy

If electrodes are placed in contact with the scalp and connected to sensitive amplifying apparatus, electrical activity originating in the brain is detectable. This activity arises from the simultaneous changes of potential in millions of nerve cells in the cerebral cortex, and it constitutes the electroencephalogram (EEG). With practice, different waves can be discerned in the rather irregular EEG tracing. The basic normal activity is the alpha rhythm, with waves occurring 8 to 13 times a second. In children it is not well marked, but in adults it is often very evident, particularly in recordings made from the back of the head; it increases when the eyes are shut and decreases when they are opened or when a mental effort is made. Beta waves (13 to 20 per second) and theta waves (4 to 7 per second) are also found in the normal EEG, but slow delta activity is always an indication of brain disease.

Epilepsy is a term which covers a variety of fits, ranging from severe convulsions to short-lived blackouts or loss of awareness. In generalized convulsions consciousness is lost, the arms and legs stiffen and then jerk rhythmically, the tongue may be bitten and the bladder emptied. The mildest fits often appear to be no more than a brief interruption of attention. Between these extremes are focal seizures (where one part of the body is involved), 'Jacksonian' fits (in which involuntary movement spreads from a thumb or the corner of the mouth to other regions) and temporal lobe attacks. Many fits begin with a premonition or aura, but temporal lobe attacks start with such symptoms as an unpleasant taste or smell, flashing lights, ringing bells, or a complete vision.

All fits are due to episodes of excessive electrical activity of the grey matter of the brain; different sorts of fit can be explained in terms of involvement of different parts of the cerebrum. Fits are in some way a constitutional upset, and tend to run in families; they can occur without obvious cause or be brought on by brain disease. Whatever their cause, they produce an abnormal EEG tracing, which may contain characteristic spiky waves during or between attacks.

An EEG machine in use

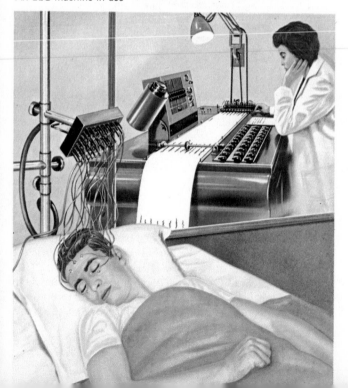

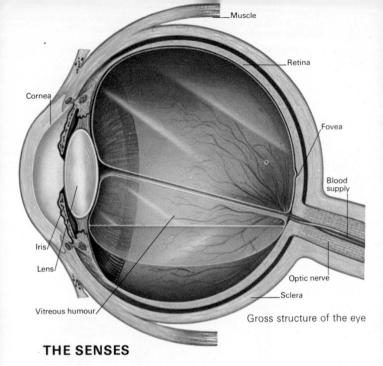

Gross structure of the eye

Labels: Muscle, Retina, Cornea, Fovea, Blood supply, Iris, Optic nerve, Lens, Sclera, Vitreous humour

THE SENSES

The eye

Like the camera, the eye collects and focuses light on to a sensitive surface. Its transparent, yellowish, gelatinous lens is much more spherical when removed than when it is in its normal position and is flattened by the pull of ligaments on its fibrous covering. Relaxation of these ligaments is produced by contraction of the ciliary muscle at the back of the iris – this allows the lens to bulge and makes nearby objects come into focus. In old age the lens becomes more rigid, and long-sightedness results.

The pupil is the aperture of the iris, which is an automatic diaphragm opening and closing in response to changes of lighting. Behind the lens the eyeball is filled with a sticky, jelly-like substance, at the back of which is the retina. This is the natural counterpart of a photographic film, and it is made up of three layers of nerve cells with the light receptors – the rods and cones – on the outside. The rods and cones are

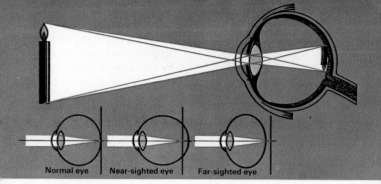

Normal eye | Near-sighted eye | Far-sighted eye

the extensions of small neurons, and they convert light into nerve impulses which are passed to the brain along the fibres of the optic nerve. Where the optic nerve leaves the eye there are no receptors: this "blind spot" is slightly off the axis of vision. The centre of the retina possesses only cones connected to nerve fibres in a 1:1 ratio. The proportion of rods increases away from the centre and receptors converge onto fewer fibres. Since rods work at low light intensity (they contain an easily bleached pigment – rhodopsin – derived from vitamin A) the periphery of the retina is adapted for sensitivity in dim light, the centre for detailed, daylight vision.

Objects to the left are seen by both eyes but projected to the right visual cortex *(middle)* Organization of receptor and nerve cells in retina *(bottom)*

An inverted image is focused on the retina. In near sight the image falls before the retina, while in far sight the image is focused behind the retina. *(above)*

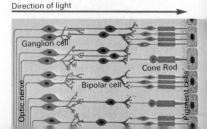

Nasal visual field

Temporal visual field

Left eye

Right eye

Optic chiasma

Optic tract

Optic nerve

Visual cortex

Visual pathway

Direction of light

Ganglion cell

Cone Rod

Optic nerve

Bipolar cell

Pigment cells

Colour vision depends upon the mixing of the messages from three types of cone corresponding to the three primary colours.

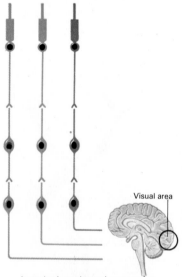

Visual area

A typical card used to test for colour blindness

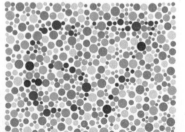

Nerve impulses pass back to the lateral geniculate body in the thalamus, whence they are relayed to the occipital cortex, and to the superior colliculus, which is concerned with control of eye movement. Fibres from the outer half of the retina stay on the same side while those from the inner half cross the midline. Since the image formed on the retina is upside down, each half-field of vision, left or right, projects on to the opposite cortex.

We see colours by virtue of possessing retinal cones sensitive to either red or blue or green, the three primary colours from mixtures of which all other colours can be produced. Thus receptors of the three types are stimulated according to the proportion of each primary in the light falling on them. Colour blindness is of various sorts, but in a common variety red cannot be recognized and probably the red sensitive cones are defective. To see colours perfectly depends on the cortex, which 'corrects' the imperfect, blurred and colour-distorted images produced by the primitive lens of the eye, just as it constructs a 'solid' image from the slightly dissimilar ones from each eye.

The ear and the mechanics of hearing

There is an outer, a middle and an inner ear. The outer ear funnels sounds on to the eardrum making it vibrate. The vibrations are transmitted to the three bones of the middle ear, the *malleus* (hammer), *incus* (anvil) and *stapes* (stirrup), which are linked to each other, to the drum on the outside and the oval window of the *cochlea* (inner ear) on the inside. This oval window has only 1/16 of the area of the ear drum and the three bones act as a mechanical linkage to concentrate the sound vibrations on to it. The cochlea is placed deep in the bone of the side of the skull. Cochlea means snail, and it is in fact a fluid-filled tube coiled $2\frac{1}{2}$ times around a central core of bone. It connects with the middle ear at the oval and round windows, covered respectively with the bony foot of the stapes and with a membrane, and it contains the actual organ of hearing, the organ of Corti, which is a collection of specialized sensory nerve processes resting on a centrally placed layer of fibres, the *basilar membrane*.

Structures of the inner, middle and outer ear

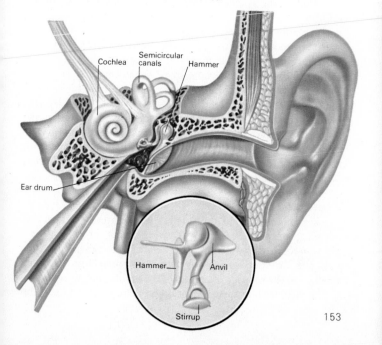

153

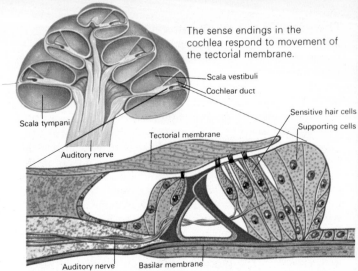

The sense endings in the cochlea respond to movement of the tectorial membrane.

Scala vestibuli
Cochlear duct
Sensitive hair cells
Supporting cells
Scala tympani
Tectorial membrane
Auditory nerve
Auditory nerve   Basilar membrane

The vibrations transmitted from the outer ear to the foot of the stapes cause corresponding minute pressure changes in the fluid of the cochlea. These pressure waves return to the middle ear at the round window after spiralling around the cochlea and activating the organ of Corti. What happens is that the sounds (or pressure waves) cause a travelling wave to appear in the basilar membrane, like the wave travelling along a loosely held skipping rope when one end is jerked. The size and speed of this wave depends on how high or low the sound is, and how loud it is. Movement of the underlying membrane stimulates the receptors on top to fire off nerve impulses. Different parts of the cochlea are sensitive to different sounds, high and low, and this difference is such that the wave produced by a particular sound excites the appropriate part of the organ of Corti.

Taste and smell

The sensations of taste and smell depend on the physical and chemical properties of the substances that produce them. They involve the attachment of small quantities of the substance in question – no more than a few molecules for some smells – to the sensory receptors of the mouth, palate and nose.

The receptors for taste are found on the upper surface of the

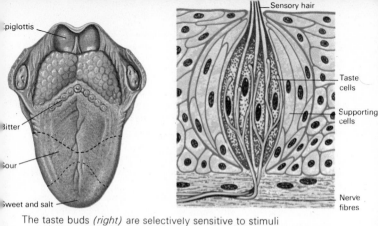

The taste buds *(right)* are selectively sensitive to stimuli and are unevenly distributed over the tongue.

tongue and in the palate and throat. The taste buds are probably of four kinds, since four basic tastes – sweet, sour, salt and bitter – are recognizable in different regions of the tongue, but they are all the same structurally, and consist of a nest of elongated cells connected to a sensory nerve. Messages initiated here are relayed to the parietal and temporal cortex.

The nerves of smell, the olfactory nerves, lie in a yellow area lining the uppermost part of the nose cavity, and are the extensions of nerve cells whose central branches pass upwards through the bone of the skull to join the olfactory bulbs. There are situated below the frontal lobes, and contain more nerve cells, the axons of which send on the sensory messages to the innermost gyri of the temporal lobes.

Endings of smell are situated in the nasal cavity.

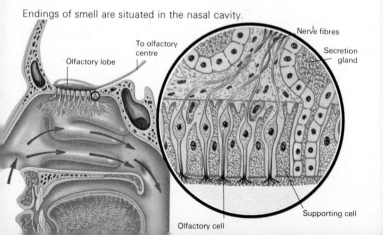

BOOKS TO READ

There are many text-books providing detailed information about Anatomy, Physiology and Medicine. Since these subjects are taught at an advanced level, or else touched upon in elementary courses of Biology and Zoology, there is a dearth of intermediate material. The following selection attempts to include some less daunting but nevertheless helpful books for the interested reader.

The Human Body by C. H. Best and N. B. Taylor. 4th Edition, Chapman and Hall, London, 1966. This is a more expansive text, but at a fairly elementary level.

Animal Physiology by Knut Schmidt-Nielsen. 2nd Edition, Prentice-Hall, New York, 1964. This short paperback forms an excellent introduction to some quantitative ideas in Physiology.

Applied Physiology by Samson Wright. 11th Edition, Oxford University Press, 1965. Popular with first year medical students, it is well diagrammed and gives a clear presentation of salient facts.

Histology by A. W. Ham and T. S. Leeson. 5th Edition, Pitman Medical Publishing Company, 1965. This is a sound general account, with many photographs, of the microscopic structure of the body's tissues.

Anatomy by Henry Gray. 34th Edition, Longmans, 1967. Generations of Anatomy students have been brought up on this standard work.

An Illustrated History of Medicine by Robert Margotta, edited by Paul Lewis, Hamlyn, 1966. This attractively illustrated book covers the development of medical treatment from the earliest times to the present.

Places to visit

Some medical schools and teaching hospitals have museums of Anatomy and Pathology. For those interested in the techniques of Surgery and Medicine there are the Wellcome Foundation museums and library.

Museum of Medical Science
Historical Medical Museum
Historical Medical Library
The Welcome Building, Euston Road, London, N.W.1.

INDEX

SOME OTHER TITLES IN THIS SERIES